PACIFIC PROFILES

VOLUME 18

Allied Bombers:
RAAF Beauforts and Beaufighters
1942–1945

KEVIN GOGLER

Illustrations by Michael Claringbould

Avonmore Books

Pacific Profiles Volume 18

Allied Bombers: RAAF Beauforts and Beaufighters 1942–1945

Kevin Gogler

Illustrations by Michael John Claringbould

ISBN: 9780975642382

First published 2025 by Avonmore Books
Avonmore Books
PO Box 217
Kent Town
South Australia 5071
Australia

Phone: (61 8) 8431 9780
avonmorebooks.com.au

A catalogue record for this book is available from the National Library of Australia

Cover design & layout by Diane Bricknell

Front Cover: A wide range of colour schemes attended RAAF Beaufighter and Beauforts which served throughout the Pacific and Australia. From top to bottom are: Beaufort A9-46 (previously RAF serial T9598) which served with No. 100 Squadron (Profile 1), Beaufort A9-548 which served with No. 15 Squadron (Profile 41); No. 30 Squadron Beaufighter A19-205 (Profile 49); No. 93 Squadron Beaufighter Mk XXI A8-116 (Profile 67); and postwar Beaufighter Mk XXI A8-265 converted to a target tug (Profile 88).

Back Cover: A No. 22 Squadron Beaufighter armed with rockets cruises over Ceram in early 1945. All Beaufighters operated by the squadron were Mk XXI aircraft painted in overall Foliage Green.

Contents

Dedication 4

About the Author 5

Glossary & Abbreviations 6

Maps 7

Introduction 11

Chapter 1 RAAF Beaufort And Beaufighter Overview 13

Chapter 2 Building the Beaufort and Beaufighter in Australia 21

Chapter 3 Camouflage and Markings 27

Chapter 4 No. 100 Squadron 35

Chapter 5 No. 7 Squadron 41

Chapter 6 No. 14 Squadron 47

Chapter 7 No. 6 Squadron 51

Chapter 8 No. 8 Squadron 55

Chapter 9 Eastern Australia Beaufort Squadrons 61

Chapter 10 Northern Territory Beaufort Squadrons 67

Chapter 11 No. 15 Squadron 71

Chapter 12 No. 30 Squadron 75

Chapter 13 No. 31 Squadron 83

Chapter 14 No. 22 Squadron 91

Chapter 15 The Late Beaufighter Squadrons 95

Chapter 16 Training Units 101

Chapter 17 Other Units 109

Chapter 18 Postwar Service 113

Chapter 19 Disposals and Survivors 117

Sources and Acknowledgements 119

Index of Names 120

Dedication

Air Commodore Keith Parsons, CBE, DSO, DFC, AFC (Rtd)

1914 – 2011

This volume is dedicated to Air Commodore Keith Parsons, CBE, DSO, DFC, AFC (Rtd), an outstanding leader of men who learnt his trade in the pre-war RAAF. Keith was the commanding officer of No. 7 Squadron from November 1942 until February 1944. As the second RAAF squadron to operate Beauforts, his sheer tenacity of purpose and hard work brought the squadron to a high state of efficiency. The welfare of his men was always paramount, and the superb spirit enjoyed by No. 7 Squadron during this period was a tribute to Keith's courage, determination and devotion to duty. Keith flew more than 100 missions and 800 operational hours with the squadron and was awarded the Distinguished Flying Cross.

Keith had joined the RAAF in January 1935 as an air cadet and in late 1939 he became the chief flying instructor of No. 1 Elementary Flying Training School at Parafield. After operational service with No. 7 Squadron, he was posted to Bomber Command in the UK where he commanded No. 460 Squadron flying Lancasters and then RAF Station Binbrook. He remained in the RAAF postwar where, amongst other postings, he commanded No. 82 Wing, the RAAF Staff College, RAAF Point Cook and RAAF Butterworth in Malaysia. He died in Adelaide in 2011, just shy of his 97th birthday.

About the Author

Kevin Gogler has been interested in Australian military aviation from an early age. Both his father and uncle were RAAF pilots, with his father flying Beauforts on operations in Australia and New Guinea with Nos. 7 and 100 Squadrons.

Kevin has been involved with several aviation museums and veterans' organisations over the years and is a member of the Air Force Association, Mitcham Branch. He has written several articles for various publications, and this is his fourth book on RAAF history. The other three books are:

Eric Gogler in the RAAF: One Pilot's War in Australia and New Guinea, self-published, 2009.

The RAAF in the Australian Coronation Contingents of 1937 and 1953, self-published, 2018.

We Never Disappoint: A History of No. 7 Squadron RAAF, 1940 to 1945, Air Power Development Centre, Canberra ACT, 2012.

Kevin Gogler in the cockpit of the Beaufort A9-557 at the Australian War Memorial in March 2006.

Glossary & Abbreviations

CSIR	Council for Scientific and Industrial Research
DAP	Department of Aircraft Production
DDT	Dichloro-diphenyl-trichloroethane, the first of the modern synthetic insecticides developed in the 1940s.
HMAS	His Majesty's Australian Ship
HP	Horse Power
NSW	New South Wales
OTU	Operational Training Unit
POW	Prisoner of War
PT	Patrol Torpedo (boat)
RAAF	Royal Australian Air Force
RAF	Royal Air Force
RAN	Royal Australian Navy
UK	United Kingdom
US	United States
VIP	Very Important Person

Explanatory Notes

Place names are, where possible, consistent with wartime usage, and with local spelling conventions.

Measurements are also consistent with wartime usage: generally, miles and miles per hour are used; altitude is given in feet.

As with many other aircraft profiles in this series, the Beauforts and Beaufighters are depicted here after taking into account ambient and sun light as well as weathering to emulate field conditions.

Note that the undersurfaces illustrated in Sky Blue, a very pale shade of blue, soon faded to an off-white and this is reflected in the variation of this colour in the profiles and to a lesser extent in the photos.

Maps

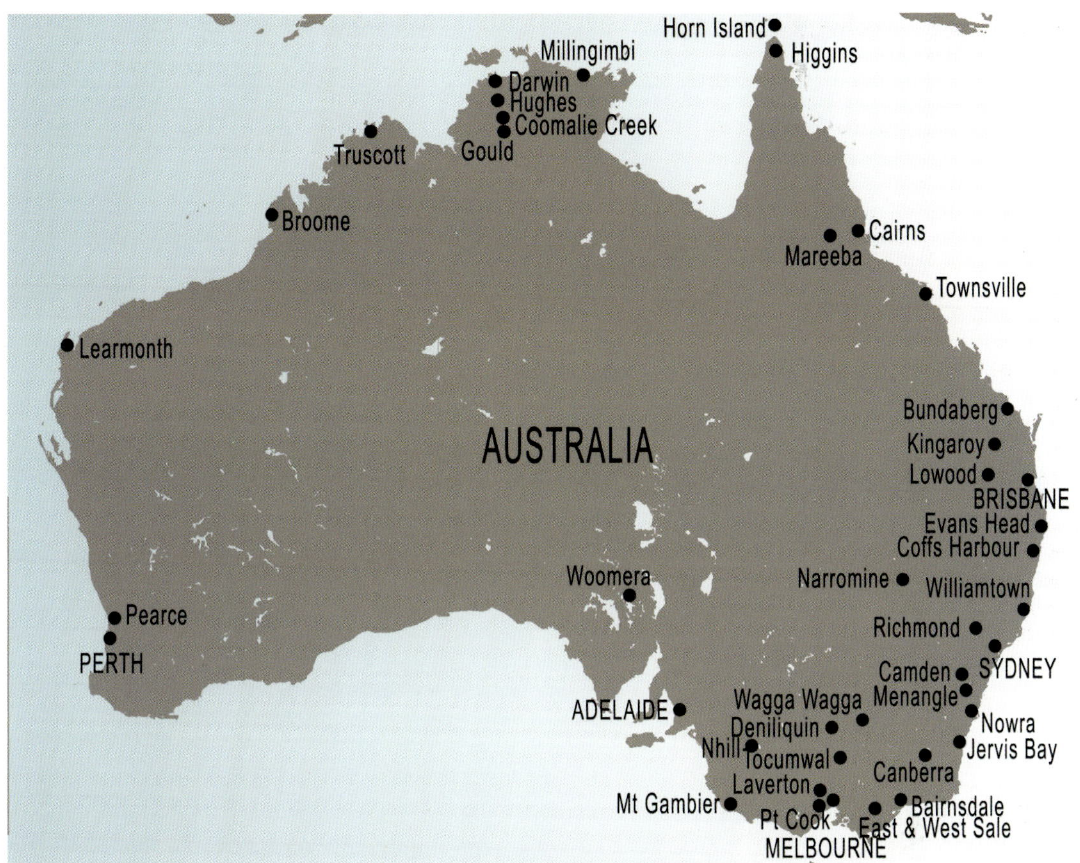

A map of mainland Australia showing most of the locations used by RAAF Beauforts and Beaufighters mentioned in this volume. Coastal airfields were used by Beaufort squadrons to conduct seaward reconnaissance missions, while several of the inland locations were used by training units or for postwar aircraft storage.

A map of wartime New Guinea with locations marked in black that were used by RAAF Beaufort and Beaufighter squadrons. In late 1942 these were limited to Port Moresby and Milne Bay, but additional airfields were captured or constructed as Allied forces generally moved northwards throughout 1943 and 1944. The three locations marked in red were important Japanese bases, although Gasmata was captured in 1944. RAAF squadrons increasingly targeted New Britain from late 1943 using island bases at Vivigani and Kiriwina.

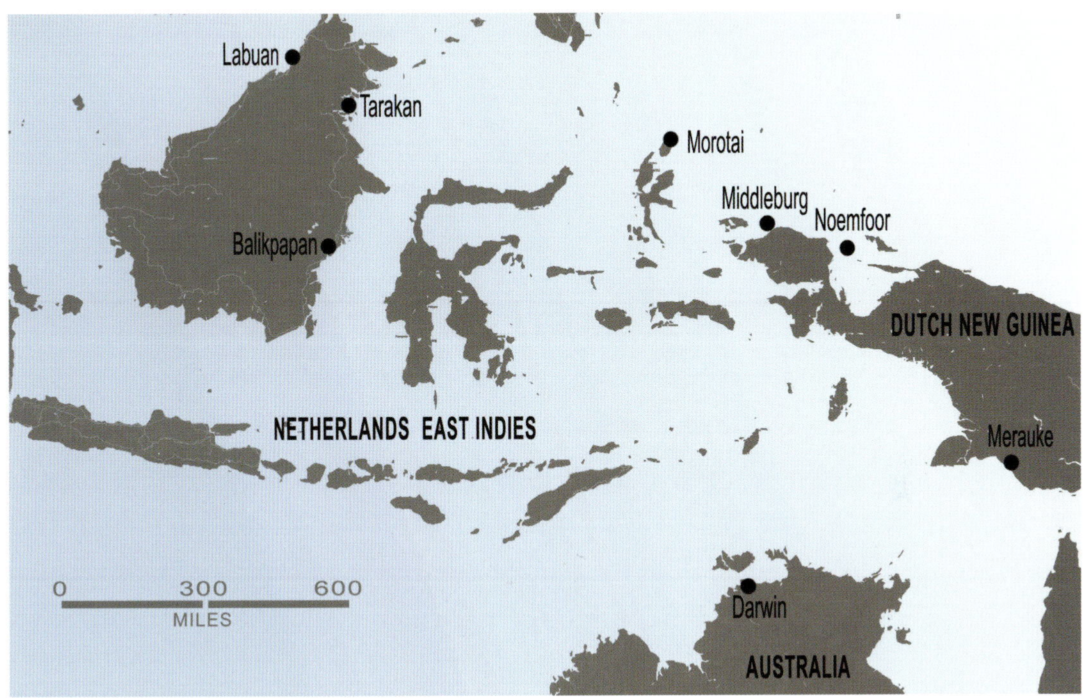

A wartime map of Dutch New Guinea and the Netherlands East Indies. From the Darwin area RAAF Beaufighter and Beaufort squadrons struck Japanese targets across the southern fringes of the Netherlands East Indies throughout the war. Further east, Merauke in Dutch New Guinea was used by No. 7 Squadron Beauforts for several months in 1943-1944.

In the final period of war, the locations marked in the northern Netherlands East Indies became operating locations for Beauforts and Beaufighters. Morotai in particular was developed into a sizeable Allied base. From mid-1945 Australian forces landed in Borneo and RAAF squadrons operated from Labuan, Tarakan and Balikpapan for a short time until the cessation of hostilities.

A fine study of a new looking Beaufort over Sydney, likely on a test flight shortly after it was built. The aircraft is A9-700, the very last Beaufort to leave the Australian production line.

A close-up of two hard worked Beauforts serving with No. 1 Operational Training Unit, circa 1943 (see Chapter 16).

Introduction

Amidst the enormous numbers of American aircraft that fought in the Pacific War, it is often forgotten that two British designed types served operationally with the RAAF in large numbers. A combined total of some 1,280 Beauforts and Beaufighters were delivered to the RAAF and served with fifteen wartime squadrons plus a variety of training and support units.

From late 1942 both types were on the frontlines of the war with Japan. The Beaufort initially had a niche as a torpedo bomber, a role not otherwise undertaken by US Fifth Air Force squadrons in the New Guinea theatre. The Beaufighter found its own niche as a heavily armed low-level attack fighter and became a key component of Allied strike missions. Both types played important roles in the landmark March 1943 Battle of the Bismarck Sea.

Subsequently the Beaufort made its mark throughout Australia, New Guinea and the Netherlands East Indies as a light bomber, a maritime patrol aircraft and by undertaking numerous support activities. The Beaufighter also served across these theatres, often undertaking coastal sweeps to hunt Japanese barges. Later versions introduced the use of rocket projectiles by the RAAF.

Overall, the contributions of Beauforts and Beaufighters to the RAAF war effort was immense. Unfortunately, some 600 young men lost their lives flying them both in combat and due to a myriad of accidents, many in Australian training units. Beaufighters also had a postwar role, and perhaps surprisingly the last was not retired from service until 1957.

Also remarkable was the fact that most Beauforts and Beaufighters were Australian produced, an exceptional technical and industrial achievement that can only be lightly touched on in a book of this format.

Note that all the profiles herein are based on photographic evidence and are reasonably true and accurate representations of the aircraft, given limitations of surviving images and documentation.

I trust this volume of *Pacific Profiles* provides an accurate overview of wartime RAAF Beaufort and Beaufighter operations and brings new life to their story from so many decades ago.

Kevin Gogler
Adelaide, South Australia
March 2025

Australian-built Beaufort A9-261 in flight. This is a Mark VIII which became the most common version and entered production from November 1942.

Australian-built Beaufighter A8-99 which was delivered in December 1944. UK built Beaufighters in RAAF service had a different "A19" serial prefix.

CHAPTER 1
RAAF Beaufort And Beaufighter Overview

The Bristol Beaufort was a twin engine light bomber, carrying a crew of four, designed for torpedo attack, aerial bombing and general reconnaissance missions. It was powered by two Pratt and Whitney 1,200hp Twin Wasp radial engines and could carry up to 2,000 pounds of bombs or one torpedo. It was armed with up to ten 0.303-inch calibre machine guns: two in the nose, one in each wing, two in the dorsal turret, one in a nose ventral cupola, one in each beam position and one in the top fuselage. RAAF Beauforts were manufactured in Australia.

Prior to the start of the Pacific War, the RAAF intended to use the Beaufort for coastal reconnaissance and the protection of shipping. In conjunction with Allied naval forces, these taskings were to be concentrated around the key ports and shipping lanes of Darwin, Albany to Fremantle, Bass Strait, Melbourne to Sydney to Newcastle and in New Guinea. The taskings entailed seaward reconnaissance for the detection of raiders and submarines, and the direct attack on any enemy vessels found. Throughout the war Beauforts were used extensively for anti-submarine patrols, offensive coastal sweeps and on routine convoy escort duties. The normal depth of these searches was to about 300 miles from the coast.

However, as Japanese forces encroached on Australia's northern borders additional use was made of the Beaufort. As the air forces in the New Guinea theatre turned to the offensive from late 1942, Beauforts were initially used in the long-range torpedo bomber role to attack Japanese shipping. However, torpedo operations were problematic and had ceased altogether by early 1944. Meanwhile Beauforts found a useful role as light bombers, attacking ships and various land installations including supply bases, fuel dumps and airstrips.

As the war progressed into 1944 and 1945, Beauforts were used increasingly in the direct support of ground forces with bombing and strafing attacks on dug-in enemy positions close to friendly forces. Such missions required the development of techniques to minimise friendly fire casualties. In due course the Beaufort squadrons came to excel at pinpoint bombing to support the New Guinea land campaign to such an extent that their air to ground operations were recognised as a model example of army/air cooperation.

To enable the concentration of effort, Beaufort squadrons in New Guinea were grouped together, with three squadrons based on Goodenough Island from the end of 1943 to the middle of 1944. From October 1944 a move to Tadji took place where Beauforts operated until the end of 1945. Beauforts also found uses in training and other support operations such as the supply of essential provisions to front line forces, urgent travel and communications flights, research and development and as a tactical transport.

Eventually some 700 Beauforts were used by the RAAF including 46 that were converted for specialised light transport duties. The six different marks of Australian Beauforts were:

Mark	**Serial Nos.**	**Remarks**
Mk V	A9-1 to A9-50	Standard Australian Beaufort airframe, S3C4-G engines with Curtiss Electric Airscrews and Mk IE turret
Mk VI	A9-51 to A9-90	Standard Australian Beaufort airframe, S1C3-G engines with Curtiss Electric Airscrews and Mk IE turret
Mk VII	A9-91 to A9-150	Standard Australian Beaufort airframe, S1C3-G engines with DH3E50 Airscrews and Mk IE turret
Mk VA	A9-151 to A9-180	Standard Australian Beaufort airframe, S3C4-G engines with Hamilton Constant Speed or DH3E50 Airscrews and Mk IE turret
Mk VIII	A9-181 to A9-529	Standard Australian Beaufort airframe, S3C4-G engines with Curtiss Electric Airscrews and Blenheim Mk V turret
	A9-530 to A9-700	Standard Australian Beaufort airframe, S3C4-G engines with Curtiss Electric Airscrews and DAP Mk VE turret
Mk IX	A9-701 to A9-746	Airframe of existing aircraft modified as a six seat, unarmed communications and freight/personnel transport minus turret, which was faired over.

The Bristol Beaufighter was a development of the Beaufort, utilising the same wing and other components but with a different fuselage and engines. About 40% of RAAF Beaufighters were imported from the UK and the remainder were manufactured in Australia. In RAAF service, the Beaufighter was a two seat, twin-engine aircraft used in the intruder cum strike/ground attack role.

The Beaufighter was powered by two Bristol Hercules radial engines of various horsepower ratings. The bomb load for the UK-built aircraft ranged from 500 pounds to 2,000 pounds depending on the version, with four 20mm cannon in the fuselage and six 0.303-inch calibre machine guns in the wings. The Australian-built aircraft also had four 20mm cannon in the fuselage but differed with four 0.5-inch calibre machine guns in the wings and the provision to mount eight 60-pound rockets.

The four different UK-built marks operated by the RAAF were:

Mark	**Serial Nos.**	**Remarks**
Mk Ic	A19-1 to A19-72	Standard UK Beaufighter airframe, Hercules XI engines with de Havilland constant-speed airscrews. Developed for Coastal Command duties with increased fuel capacity. Drum ammunition for cannons.
Mk VIc	A19-73 to A19-136	Standard UK Beaufighter airframe, Hercules VI or XVI engines with de Havilland constant-speed airscrews. Some with dihedral tailplane. Belted ammunition for cannons.

Mark	**Serial Nos.**	**Remarks**
Mk XIc	A19-137 to A19-148 A19-151 & 152 A19-156 to A19-158 A19-160 & 161 A19-163	Standard UK Beaufighter airframe, Hercules XVII engines with de Havilland constant-speed airscrews. Dihedral tailplane and belted ammunition for cannons.
Mk X	A19-149 & 150 A19-153 to A19-155 A19-159 & 162 A19-164 to A19-218	Standard UK Beaufighter airframe, Hercules XVIII engines with de Havilland constant-speed airscrews. Dihedral tailplane and belted ammunition for cannons.

There was only one Australian-built type of Beaufighter, the Mk XXI:

Mark No.	**Serial Nos.**	**Remarks**
Mk XXI	A8-1 to A8-365	Standard Australian Beaufighter airframe, Hercules XVIII engines with de Havilland hydromatic 55/14 constant-speed airscrews. Belted ammunition for cannons.

The Beaufighter was initially ordered as a long range, two-seat fighter for the RAAF. However, the type quickly found a natural niche in conducting low-level attacks against Japanese shipping and land targets. With a heavy armament of cannons and machine guns and the ability to carry bombs and rocket projectiles, Beaufighters were used throughout New Guinea and the Netherlands East Indies in attacks on shipping, airfields, seaplane bases, enemy held villages, troop concentrations and supply dumps. It was not until the beginning of 1945 that the Beaufighter squadrons were based together, firstly at Morotai and then at Tarakan.

With the ready availability of Beaufighters from late 1944, it was proposed to replace several squadrons of Beauforts in front line service with Beaufighters, but this never eventuated. Like the Beaufort, the Beaufighter was successfully used for training and in various other support operations, such as urgent travel and communications activities and research and development.

For training and refresher flying, Beauforts were fitted with dual controls, and each Beaufort and Beaufighter squadron had one or two Beauforts so fitted. Following the end of hostilities in 1945, the Beaufighter continued to be operated by the RAAF for a further ten years in the target towing, technical training, air trials and air navigation roles. Some 580 Beaufighters were used by the RAAF.

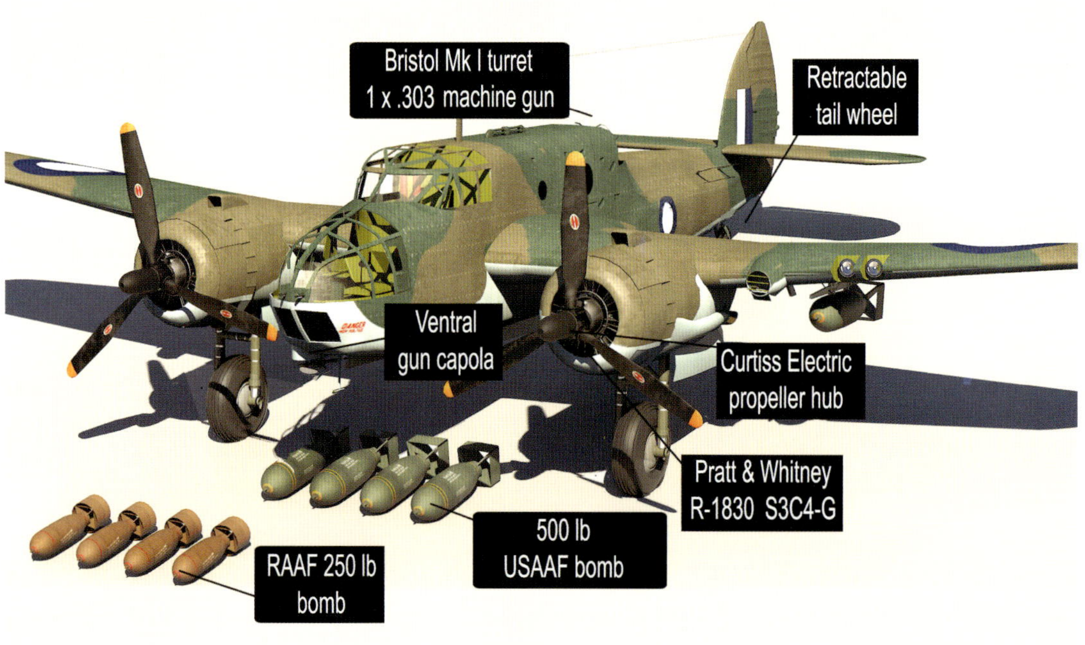

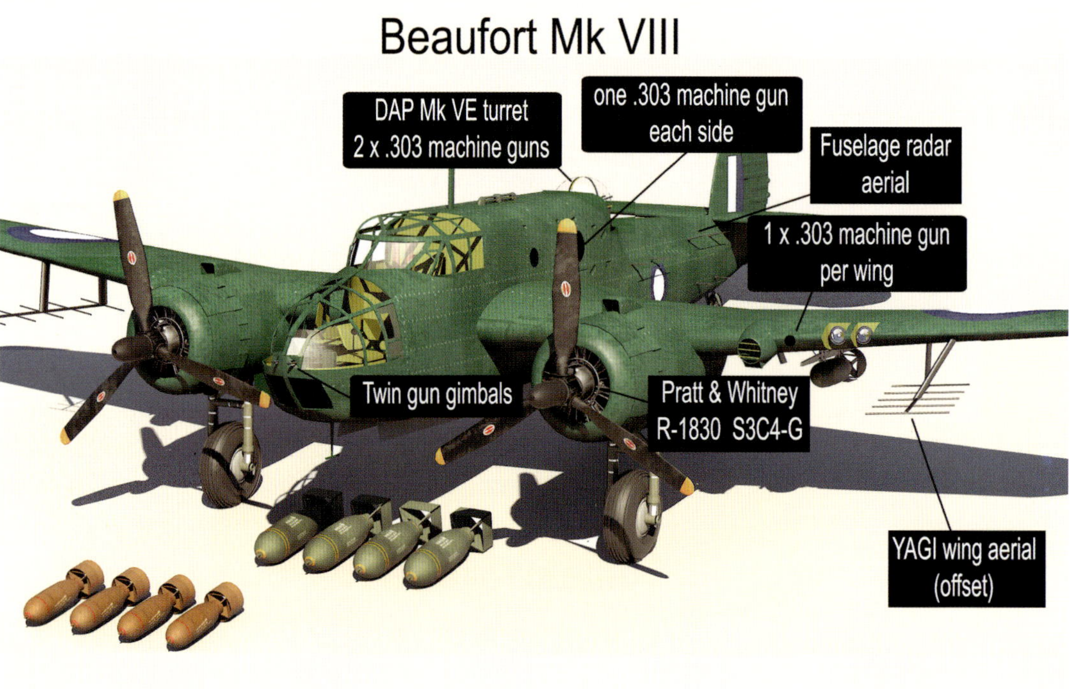

A comparison of the armaments of the Beaufort Mk V and Mk VIII. The Mk V was the initial production version, but the Mk VIII was the main production version and had a heavier machine gun armament as well as radar.

A three-view illustration showing the layout of the Beaufort Mk VIII.

A comparison of the armaments of the Beaufighter Mk Ic and Mk XXI. The Mk Ic was the first UK-built variant operated by the RAAF. The later Australian-built Mk XXI had more powerful engines and a much heavier armament including under wing rockets.

A three-view illustration showing the layout of the Beaufighter Mk Ic.

The sole UK-built Beaufort to be delivered to Australia, L4448. It was imported as the pattern aircraft for Australian production and is shown here about the middle of 1941 with the UK-style fin, which was later increased in size for improved stability. The aircraft was later allocated RAAF serial A9-1001 and entered service with No. 4 Communication Flight. It crashed on take-off from Darwin in February 1943 and was subsequently converted to components.

Built at Bristol's Weston-super-Mare shadow factory, Beaufighter LX815 was a Mk X delivered as a pattern aircraft for Australian Beaufighter production. It arrived in Australia in September 1943 and is shown here about the middle of 1944 in the RAF Coastal Command colour scheme of white with Extra Dark Sea Grey top surfaces. Allocated RAAF serial A19-218, it served only briefly from August 1945 before it was stored in 1946 and struck off charge in 1949.

CHAPTER 2

Building the Beaufort and Beaufighter in Australia

In the 1930s the RAAF was considering land-based general reconnaissance aircraft that could operate off Australia's coasts and replace the Avro Anson in the general reconnaissance role. Aircraft were needed in relatively large numbers to cover shipping routes, key ports and the widely dispersed Australian capital cities.

The long nose variant of the Bristol Blenheim, the Bolingbroke, was considered best suited to Australia's needs, and an order for 40 was placed in February 1937. However, within a year the RAAF was considering replacing the Bolingbroke with the Bristol Beaufort. The Beaufort was a further improvement on the Bolingbroke, as it combined the general reconnaissance, torpedo bombing and general-purpose functions in one aircraft.

Hence in March 1938 the order for the Bolingbroke was changed to the Beaufort. Just five months later a further 40 Beauforts were added, with the option that an additional 47 be ordered when further funds became available. All these aircraft were to be powered by the Bristol Taurus engine, an untried sleeve-valve engine. The RAAF was concerned to guard against a breakdown in the supply of the Beaufort by providing for their local manufacture. The contract for the Beaufort, therefore, included a clause covering the manufacture of both the airframe and engines in Australia.

The projected delivery date for the Beauforts proved to be extremely optimistic, as all UK production of the Beaufort was destined for the RAF following heightened tensions in Europe. By the time war was declared in 1939 no Beauforts had been shipped to Australia. The UK government then seemed to put a proviso on whether the Beauforts were to be sent to Australia at all; that proviso centred around what contribution Australia was prepared to make to British fighting forces during the war. A previous proposal for RAAF personnel to fly UK manufactured Beauforts to Australia was refused, as the personnel necessary to move such a number of aircraft was not available.

Indeed, various delays in UK manufacturing and flight trials had held up Beaufort deliveries and the type did not enter RAF service until early 1940. By this time RAAF attention had become focused on the manufacture of the Beaufort in Australia, and the re-equipment of several RAAF squadrons with UK-built Beauforts was severely delayed and eventually cancelled outright. Only one UK built Beaufort was delivered to Australia, and that was to be used as a pattern aircraft for local manufacture of the type.

The delivery of Beaufighters from the UK was a different story, however. As part of the Home Defence Plan to establish nineteen squadrons, the RAAF had determined that one long range fighter squadron was necessary to provide protection of the vulnerable industrial areas around Sydney from possible ship-borne enemy aircraft. The new Bristol Beaufighter was considered the most suitable type for this purpose and eighteen of these aircraft were ordered from the UK in late 1939. However, delivery did not eventuate as all Beaufighter production was needed for RAF squadrons. The RAAF reconsidered this requirement and ultimately decided general reconnaissance aircraft were a higher priority. Accordingly, the order for these eighteen Beaufighters was cancelled.

Only a few months later, the Australian government agreed to increase the number of RAAF squadrons from 19 to 32. From the end of 1940 and into 1941 aircraft were sought from the UK and the US but demand from other countries was high. The RAAF prioritised the new types it required for the additional squadrons. While long range, two-seat fighters were still required their priority was deemed to be subordinate to general reconnaissance, general purpose, fleet cooperation and dive-bomber aircraft. Nevertheless, the UK government did set aside 54 Beaufighters to equip two operational RAAF squadrons and a training unit.

After some delay the first shipment was not received until March 1942. As the first squadron was forming, the RAAF ordered a further eighteen Beaufighters to cover forecast attrition from accidents and combat. After the two new squadrons had seen action in New Guinea and from the Northern Territory, war wastage (as it was called) resulted in another order in early 1943 for 72 Beaufighters. This was followed by a repeat order in August 1943 for another 72 aircraft, by which time it was expected local production would negate the need for further deliveries from the UK. One additional Beaufighter was ordered separately, to be used as a pattern aircraft for the manufacture of the type. Of the 217 UK built Beaufighters, all were received except for seven: six were lost at sea during shipment and one crashed in the UK before it was crated for delivery.

During 1938 the RAAF had indicated its preference for some of the Bristol Beauforts on order to be manufactured in Australia but with no company in a position to do so, the proposal lapsed. At the end of that year, the Australian government agreed for a UK government Air Mission to visit Australia to investigate a scheme for local military aircraft manufacturing. The Air Mission considered such a scheme to be practical, and arrangements were put into place for 180 Beaufort airframes, and possibly engines, to be manufactured. Railway organisations and contractors in the southern states would produce various components which would then be assembled at specially built facilities, one in Sydney and the other in Melbourne. The first 90 Beauforts to be built were intended for RAF squadrons in the Far East with the remaining 90 for RAAF use.

The initial production timetable for the scheme was ambitious, with 180 airframes to be manufactured between July 1940 and October 1941. However, this was severely upset by the outbreak of war in September 1939 which delayed the supply of essential jigs and tools and, more importantly, support from the Bristol Aeroplane Company in the UK. Another major delay was the decision to fit a different engine, the US designed Pratt and Whitney Twin-Row Wasp rather than the Bristol Taurus. A further critical part of the scheme was the need to train specialist staff, with almost 100 tradesmen, foremen and inspectors sent to the UK to be trained at the Bristol factory. To further assist the Australian scheme, Bristol dispatched one complete airframe and sufficient parts to complete the first twenty aircraft. Certain major components continued to be supplied from overseas for some time until Australian production of those parts could be organised.

In the early months delivery schedules kept slipping due to the non-availability of essential components. The first production aircraft was completed in early August 1941 but only ten had been received by the end of that year. One of the big delays at this time were the engine assemblies from the Lockheed Corporation in the US. The assemblies were the same as for the Hudson aircraft, at that time already in service with the RAAF.

A major limitation on the production of the Beaufort was the availability of engines. The Twin-Row Wasps had to be imported from the US although plans were put in place for that engine to be manufactured in Australia. A purpose-built facility was constructed in Sydney but could not meet demand and engines continued to be imported until the program was completed.

The entry of Japan into the war and the subsequent threat to Australia prioritised the needed materials and a steady supply finally began to be received from local production and from overseas. Due to the immediate Australian need for as many Beauforts as possible and the reduced RAF need, the UK government agreed for its 90 contracted Beauforts to be instead delivered to the RAAF.

In July 1942 the number of Beauforts on order was increased from 270 to 450, and then only six months later a further extension was approved for a total of 700 aircraft plus spares. Production rose to a rate of about 30 aircraft per month until the last Beaufort was delivered in August 1944. However, during 1944 and 1945, 46 existing airframes from the 700 manufactured were converted into a light transport/communications version, known as the "Beaufreighter".

Considerations relating to the manufacture of the Beaufighter in Australia began about September 1941 when the Australian government was considering the future of aircraft construction and what types could be manufactured following on from the Beaufort. At this time, discussions were underway with the UK government to manufacture the Beaufighter in Australia, a proposition helped by the fact that over 70% of parts were interchangeable with the Beaufort. Initially the UK government was not supportive but about nine months later they did a complete about turn and recommended Australian manufacture of the Beaufighter. The small number of modifications required to UK-built aircraft to meet RAAF service requirements made manufacturing the aircraft in Australia viable.

In January 1943 arrangements were put in place for tooling up for Beaufighter production and the number of aircraft ordered was initially set at 350. This number was calculated as not only being of sufficient numbers to meet the operational requirements of the RAAF for that type of aircraft but also to maintain intact the Beaufort production establishment, its skilled labour force and the many sub-contractors until such time as the manufacture of a new type of aircraft was selected (ultimately the Avro Lincoln). This number was later increased by a further 100 to 450 based on the projection of when Lincoln manufacture would commence.

Production of Beaufighters was arranged so that deliveries of Beauforts decreased at approximately the same rate as Beaufighter deliveries increased. The first Beaufighter from

Australian production was delivered to the RAAF at the end of May 1944 and deliveries of Beauforts continued concurrently until August 1944 when the Beaufort program was completed. Beaufighter production then continued at approximately 30 aircraft per month. Following the cessation of hostilities and a subsequent review of RAAF future aircraft requirements, it was decided that the number of Beaufighters manufactured should terminate at approximately 360. In the end, 365 Beaufighters were delivered to the RAAF.

The engines to power the Beaufighter remained the Bristol Hercules for the Australian-built version resulting in the need for all these engines to be imported from the UK. The UK government was able to deliver the numbers required throughout the period of the program. As insurance against an interruption to supply, one prototype Beaufighter was fitted with Wright Cyclone engines imported from the US, but in the end this option was never needed.

The story of the manufacture in Australia of the Beaufort aircraft, and to a lesser extent the Beaufighter, was a magnificent achievement given the many obstacles that were encountered. At the commencement of the war in 1939, Australia had a limited industrial capacity, especially in respect to aircraft construction. There was a limited pool of technicians and skilled labour available, and there were no facilities for producing machine tools. Hence the setup for Beaufort manufacture involved not only the formation of an organisation capable of mass producing an all-metal, mid-wing monoplane with stressed skin, but the establishment and development of a large cluster of technically complex industries. Coordination of these activities was an enormous task falling under the Department of Aircraft Production (DAP), and accordingly these Australian-manufactured aircraft are often referred to as "DAP Beauforts" and "DAP Beaufighters".

While this Australian manufacturing story was a great achievement, it also witnessed an unfortunate outcome. In 1942 and 1943 a number of Beauforts disappeared or were seen to crash in unexplainable circumstances. These losses became a serious morale issue, particularly within training units. It was not until early 1944 that the problem was eventually discovered to be the elevator trim tab control cable support. Mild steel spring washers had been used rather than high tensile steel as specified. This resulted in the eventual failure of the elevator trim tab such that it flicked up or down, rather than remaining in the neutral position. The aircraft would then experience uncontrollable porpoising before a subsequent dive into the ground. It is unfortunate that the discovery of this mistake and its rectification took so long with the loss of so many lives.

Beaufort airframes under assembly at Mascot, NSW, around September 1943.

The production of Beaufighter airframes at Mascot, around September 1944.

The first 90 Australian-built Beauforts were intended for RAF squadrons in the Far East. Accordingly, as shown here, they were assembled with RAF markings and camouflage. They also had RAF serial numbers, with the closest airframe in this photo being T9626. Virtually all of these Beauforts subsequently entered RAAF service and were allocated RAAF "A9" prefix serial numbers.

A line-up of newly built Beauforts outside the Fishermans Bend factory. These have been constructed with RAAF markings and camouflage. The second aircraft is A9-203.

CHAPTER 3
Camouflage and Markings

The determination of camouflage colours for RAAF Beauforts and Beaufighters is not straightforward, although the Beaufort is a little easier. Documentation from the period is not complete and just because RAAF orders specified particular colours or schemes, it does not necessarily mean those were universally adopted. A reasonable estimation is provided below.

The first order of Beauforts manufactured in Australia were for the RAF in Malaya and the disruptive camouflage scheme requested was applicable for that location and consistent with existing RAF practice. The camouflage matched the RAF Dark Green and Dark Earth colours on the upper surfaces and were applied in two distinct camouflage schemes, one being the mirror image of the other. The undersurfaces were Sky Blue, a very pale shade of blue, which soon faded to an off-white.

In early 1942, the RAAF took over the RAF Beaufort order and aircraft continued to come off the production lines in the original colour scheme, although some aircraft seem to have been painted with a lighter brown. This continued until stocks of the original colours were replaced with comparable RAAF colours of Foliage Green and Earth Brown. All RAAF Beauforts delivered from the two assembly plants were in this scheme. In operational areas, the undersurfaces of some Beauforts were painted Night Black to suit nocturnal missions.

In May 1944, the RAAF revised its camouflage scheme so that Beauforts were to be painted overall Foliage Green. This simplified painting and provided for the best general purpose basic finish and was to be implemented for all aircraft, regardless of location or use. When this revised scheme came into force, there were only 50 or so Beauforts to be delivered before production ceased and these aircraft continued to be camouflaged with the disruptive brown and green camouflage with Sky Blue undersurfaces. The revised camouflage scheme was only applied when aircraft were required to be serviced or when being transferred to operational areas. For example, in July 1944 No. 15 Squadron was required to strip its aircraft of their Foliage Green and Earth Brown camouflage. This was replaced with an overall Foliage Green scheme prior to the unit's forthcoming move to New Guinea. Many Beauforts already serving in operational areas continued in their earlier three-colour scheme for some time.

The exact camouflage of UK-built Beaufighters is an unknown quantity to some extent. The first Beaufighters arriving from the UK were camouflaged in the relevant RAF disruptive scheme of Dark Green and Dark Earth over Sky Blue. However, later deliveries of the temperate scheme had Ocean Grey in the place of brown or were in the coastal scheme of Dark Slate Grey and Extra Dark Sea Grey over Sky. It is thought that the last few UK Beaufighters delivered were in the latest coastal scheme of Extra Dark Sea Grey over Sky. These aircraft may have been repainted if the delivery scheme was thought unsuitable for the aircraft's intended area of operations, for example using Sky Blue instead of Sky. In many cases, the aircraft were issued to squadrons in their original schemes. In addition, RAF colours deteriorated rapidly under tropical conditions and were soon replaced with similar colours or more likely with the RAAF colours of Foliage Green and Earth Brown. In July 1943 a request to allow varying camouflage schemes for different localities was rejected on the basis that the reallotment of aircraft between localities would render such a scheme unworkable.

These disruptive camouflage schemes continued until the May 1944 simplified Foliage Green scheme. At that time, the first Beaufighters to be manufactured in Australia were coming off the production lines and these were delivered in the overall Foliage Green scheme (although the first twenty or so were originally painted with Foliage Green upper surfaces and Azure Blue undersurfaces but the blue was painted out before delivery). However, by 1945 some Beaufighters at No. 5 Operational Training Unit were taken back to natural metal finish and operated as such. A small number of Beauforts and Beaufighters, mainly with communications units but one or two with operational squadrons, were also taken back to a natural metal finish.

Beaufighters retained in use postwar generally kept their overall Foliage Green for some time, even though all aircraft expected to remain in service were to have their camouflage removed. However, Beaufighters used for target towing duties were painted in yellow and black stripes all over, but this was later changed to aluminium with yellow and black stripes on the undersurfaces and a yellow band around the rear fuselage.

As previously stated, the initial Beauforts manufactured in Australia were for the RAF in Malaya and the first six were flown to Singapore from Melbourne in December 1941. On arrival, they were coded by the RAF with the code letters "NK". When these Beauforts returned to Australia later that month, they continued to operate with those codes for a short period until they were painted out.

Under the nineteen squadron Home Defence Plan of 1939, RAAF squadrons in Australia were allocated with single letter codes, starting at "A" for No. 1 Squadron and continuing down the alphabet to "V" for No. 25 Squadron. However, when Beauforts started to be issued to units in early 1942, the system was already outdated as No. 100 Squadron was formed in March 1942 and could obviously not comply with the single letter scheme. There is no evidence that Beauforts issued to squadrons in 1942 carried single letter codes. However, some Beaufighters were coded with individual single letters prior to displaying their squadron codes.

Due to the expansion of the RAAF, by the end of 1942 it had become overdue to introduce a system of code letters for all operational and reserve squadrons. A new system was introduced in January 1943 whereby letters were to be applied over the camouflage on each side of the fuselage directly forward or aft of the fuselage roundel. The code for operational squadrons took the form of two letters of the alphabet chosen by RAAF headquarters to identify each squadron, there being no sequence in the choice of the letters. Each aircraft in the squadron was also identified by an individual single letter of the alphabet, allocated at squadron level. However, the code letters "C" and "I" were not to be used (although there is photographic evidence that they were).

If there were more than 24 aircraft in a squadron, then additional individual code letters were to have a bar across the top of the letter, starting again at "A". All letters were to be the same height as the fuselage roundel and in the colour Sky Blue, later changed to medium Sea Grey. As can be expected, there were variations. Beaufort and Beaufighter equipped squadrons and units were allocated the following codes:

Squadron/Unit	Aircraft	Code
No. 1 Squadron	Beaufort	NA
No. 2 Squadron	Beaufort	KO
No. 6 Squadron	Beaufort	FX
No. 7 Squadron	Beaufort	KT
No. 8 Squadron	Beaufort	UV
No. 13 Squadron	Beaufort	SF
No. 14 Squadron	Beaufort	PN
No. 15 Squadron	Beaufort	DD
No. 22 Squadron	Beaufighter	DU
No. 30 Squadron	Beaufighter	LY
No. 31 Squadron	Beaufighter	EH
No. 32 Squadron	Beaufort	JM
No. 92 Squadron	Beaufighter	OB
No. 93 Squadron	Beaufighter	SK
No. 100 Squadron	Beaufort	QH
No. 1 Communication Unit	Beaufighter and Beaufort	EV
No. 3 Communication Unit	Beaufighter and Beaufort	DB
No. 4 Communication Unit	Beaufighter and Beaufort	VM
No. 5 Communication Unit	Beaufort	KF
No. 6 Communication Unit	Beaufort	XJ
No. 7 Communication Unit	Beaufort	YB
No. 8 Communication Unit	Beaufighter and Beaufort	ZA
No. 11 Communication Unit	Beaufort	HM
No. 9 Local Air Supply Unit	Beaufort	TX
No. 10 Local Air Supply Unit	Beaufort	UB
No. 12 Local Air Supply Unit	Beaufort	TA
No. 111 Air Sea Rescue Flight	Beaufort	KP

Many of the aircraft used by the communication and local air supply units remained uncoded and some were only coded with a single letter. The same applied to some training units, however, the Beauforts of No. 1 Operational Training Unit had the individual aircraft serial number in large yellow numbers on each side and under the nose.

The first 100 or so Beauforts manufactured in Australia were delivered from the assembly plants with the standard RAF style roundels of red surrounded by blue on the upper surfaces and red surrounded by white surrounded by blue with a yellow outer ring on the fuselage sides. Fin striping was red, white and blue. Initially there were no roundels on the under surfaces, but these were added later. In September 1942 the RAAF deleted the red circle of these roundels altogether by overpainting them with white to avoid confusion with the Japanese *hinomaru* marking. The Americans had deleted the red disc from their markings a few months earlier. The yellow fuselage circle had generally also been overpainted sometime previously. The red fin striping was also removed.

Beaufighters received from the UK retained their RAF style roundels until September 1942 when the red was similarly deleted from those aircraft. The white roundel surrounded by a blue ring remained the national marking for the rest of the war and for a period after, but the size of the white disc did reduce with a consequential increase in the size of the blue ring. This was to find a better balance between displaying national markings and compromising the camouflage of the aircraft. In 1948 the RAAF reverted to the standard RAF style roundel until 1956 when the kangaroo roundel was adopted. Therefore, only postwar Beaufighters reverted to the standard RAF style roundel with only one or two wearing the kangaroo roundel.

Like aircraft everywhere, Beauforts and Beaufighters were adorned with what is commonly called nose art, although it was not always on the nose of the aircraft. In late 1942 and early 1943 the application of nose art to Nos. 7 and 100 Squadron Beauforts was enthusiastically taken up by crews. However, this practice seems to have faded to some extent, although not completely. This was because pilots and their crews soon realised that they didn't retain their individually adorned aircraft for long, particularly when those aircraft needed to be serviced away from the squadron and different aircraft were assigned. Other Beaufort and Beaufighter squadrons did dabble in the practice, but it was not encouraged by higher authority.

In fact, RAAF Command was concerned that aircraft of the Allied air forces, engaged in combat, had various names and slogans painted on them which made excellent propaganda material for the enemy and which could bring retaliation against air force personnel shot down and captured by the enemy. A direction was issued in January 1944 that immediate steps were to be taken to ensure that nothing was painted on an aircraft, clothing or equipment that might be used as propaganda material by the enemy, or which might tend to bring action against captured personnel.

Later that year and into 1945, the nose art practice seemed to regain popularity. Beaufighter aircraft had smaller flat fuselage surfaces to apply such nose art, but names and smaller emblems were witnessed. No. 93 Squadron applied its art to both sides of the rudder. As bombing missions on Beaufort aircraft began to mount up, tally markings were added to the port side of the aircraft below the cockpit. Some aircraft eventually mounted an impressive tally. Most tally markings were a symbol of a bomb, or a parachute for a supply dropping mission. In No. 7 Squadron during 1945, when it was pointed out that that these missions were more of a "milk run", the bomb symbol was changed to a milk bottle symbol (see photo on page 40).

Illustrated are the main camouflage schemes used by operational Beauforts and Beaufighters as described in the text. From top to bottom: Beaufighter RAF Dark Green and Dark Earth; Beaufighter RAF Dark Slate Grey and Extra Dark Slate Grey; Beaufighter RAAF Foliage Green; Beaufort RAAF Foliage Green and Earth Brown; Beaufort RAAF Foliage Green.

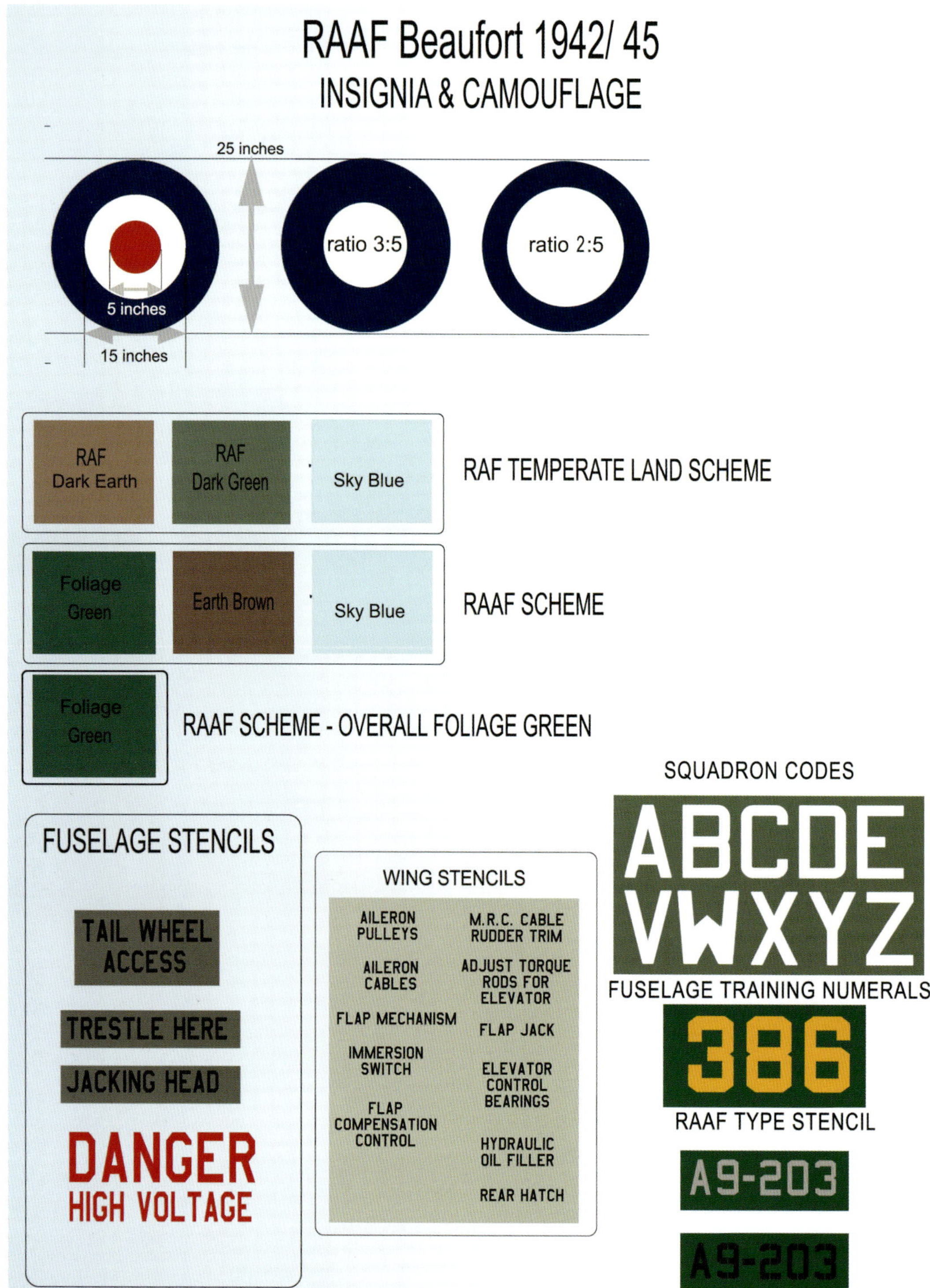

Details of markings and camouflage colours used on operational RAAF Beauforts.

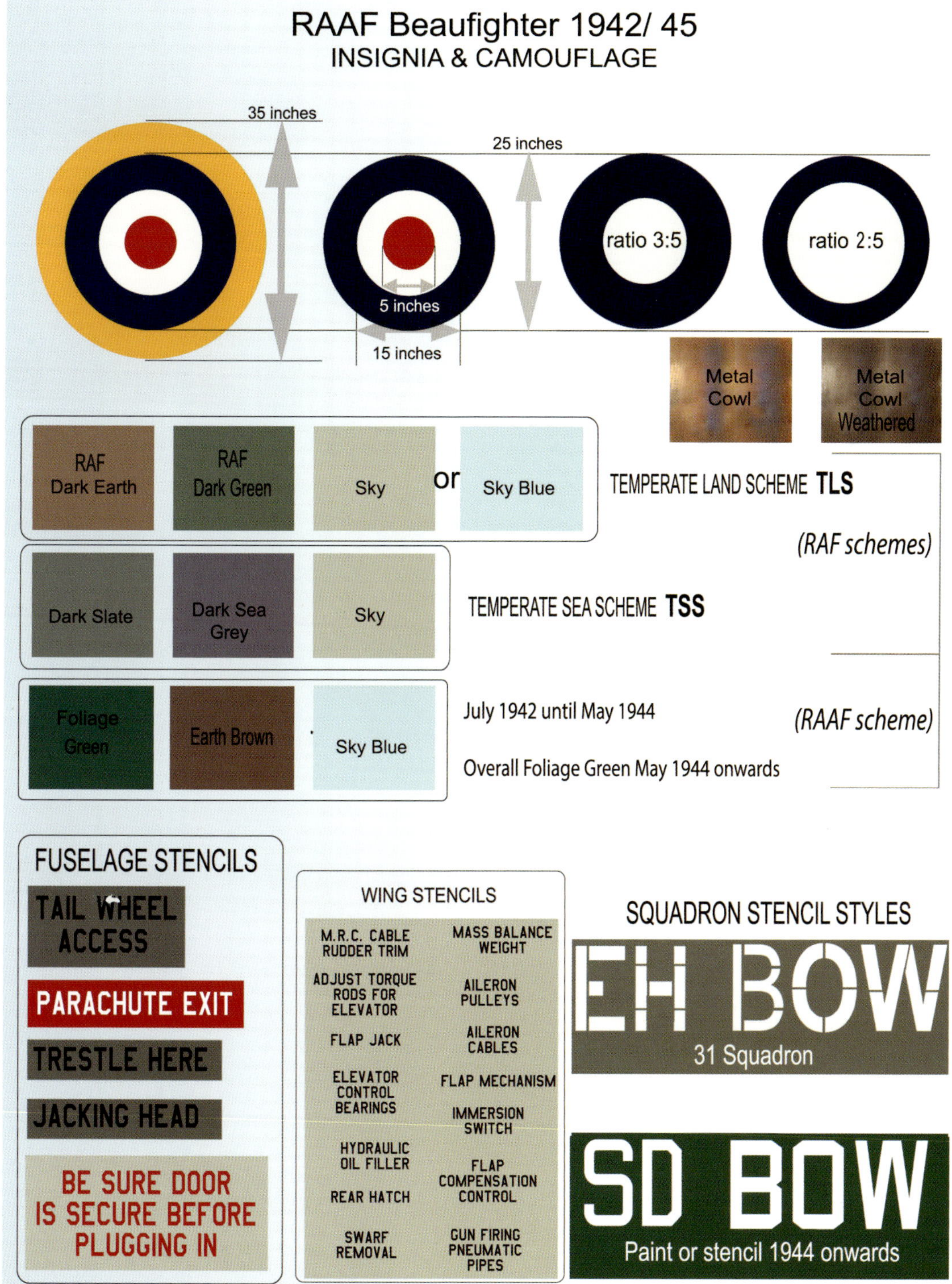

Details of markings and camouflage colours used on operational RAAF Beaufighters.

A9-54 was an early Beaufort Mk VI that served with No. 100 Squadron from June until October 1942 when it was damaged while landing with a flat tyre. It is photographed here at Coen in Far North Queensland circa June 1942. The winged dog motif on the starboard side of the nose is Flying Snifter, a contemporary cartoon character. The Malayan phrase "Kita Bikin Ayer Sama Tojo" translates as "We Piss on Tojo". The camouflage is Dark Green and Dark Earth over Sky Blue. In October 1944 the aircraft force-landed in the sea off New Guinea while serving with No. 9 Communication Unit and was written off.

CHAPTER 4
No. 100 Squadron

The first RAAF Beaufort squadron to form was No. 100 Squadron in February 1942 at Richmond, NSW. The unit absorbed the aircraft and, temporarily, the crews of the RAF squadron of the same number, at that time on loan to the RAAF and under training at Richmond. Wing Commander Sam Balmer was appointed the commanding officer and served for the next twelve months. Training soon commenced when sufficient personnel had arrived but interspersed with this training anti-submarine patrols and convoy escorts were undertaken along the east coast. Operations and training were hindered due to high unserviceability rates experienced with the new Beauforts.

In May the squadron moved first to Cairns and then to Mareeba, Queensland, where anti-submarine and seaward reconnaissance patrols continued. One aircraft was lost during this time. The first offensive operation occurred on 27 May when two aircraft bombed the Japanese floatplane base in the Deboyne Islands. Then in June, aircraft proceeded to Port Moresby to attack Salamaua and shipping off Lae at night. However, a further two aircraft were lost. At the end of the month, the squadron moved back to Laverton, Victoria, for further training. Anti-submarine patrols and convoy escorts continued during July and August from Mount Gambier in the west to Mallacoota in the east. An attack was made on an enemy submarine off Mallacoota. While these operations were underway, some crews moved to Nowra, NSW, to commence torpedo training.

From September the squadron began moving north again, first to Bohle River near Townsville and then to Milne Bay where limited torpedo operations were carried out. Attacks were made against shipping along the New Guinea coast together with reconnaissance patrols to the north and east. These operations continued for the remainder of the year with five aircraft lost in the last three months of 1942.

The new year of 1943 began with a torpedo attack on a convoy off Gasmata, but bad weather resulted in two aircraft crashing into hills on the return flight. In addition, bombing attacks, anti-submarine patrols and reconnaissance flights were a feature of the first few months of the year. Also during this time No. 100 Squadron Beauforts began to be fitted with radar and crews were trained in its use. Enemy shipping continued to be attacked with torpedoes, but these proved to be frustratingly unreliable.

Considerable training was carried out in March to improve torpedo attack methods, especially by night and with the assistance of radar. However, there were limited opportunities to put this training into practice, so operations instead consisted of large-scale night attacks on Salamaua, the strafing of barges and rafts and bombing missions against Gasmata airstrip. The squadron's torpedo operations ceased altogether at about this time, while during June and July aircrews from No. 7 Squadron joined the unit. In the largest attack for the RAAF to date, Beauforts from No. 100 Squadron, together with Bostons, Beaufighters and Kittyhawks struck Gasmata.

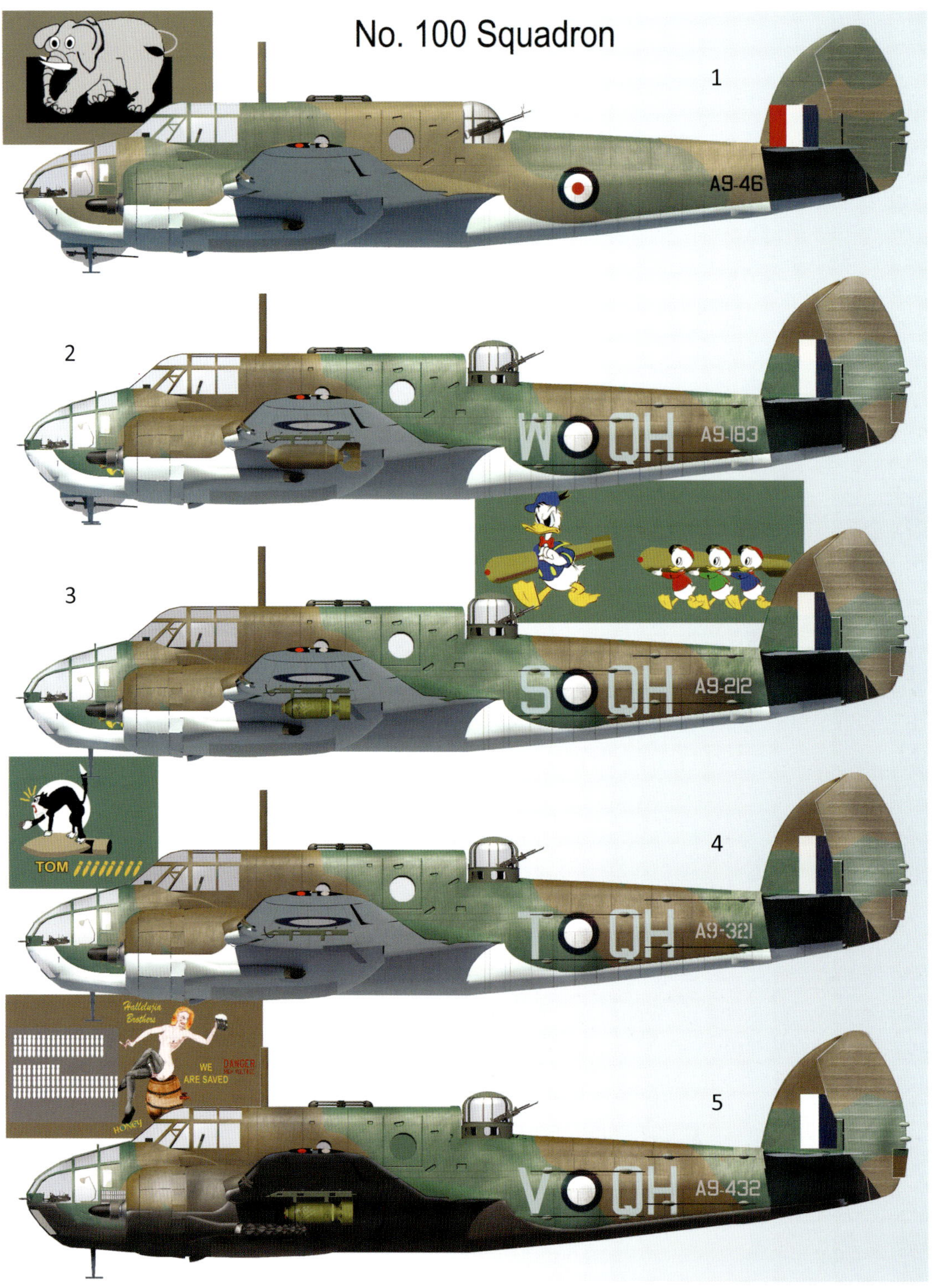
No. 100 Squadron
1
A9-46
2
W QH
A9-183
3
S QH
A9-212
TOM
4
T QH
A9-321
Hallelujah Brothers
WE ARE SAVED
DANGER
HONEY
5
V QH
A9-432

Profile 1 Beaufort Mk V A9-46

This Beaufort, previously RAF serial T9598, served with No. 100 Squadron from May 1942 until December of that year. It subsequently served for the remainder of the war in a training capacity with a variety of units. In November 1945 it was stored at Wagga Wagga, prior to being struck off charge in August 1949.

Profile 2 Beaufort Mk VIII A9-183, W-QH

This Beaufort served with No. 100 Squadron from December 1942 until it went missing during an attack on Gasmata airfield, New Britain, in September 1943. The aircraft had been hit by anti-aircraft fire and crashed nearby with all of the crew being killed.

Profile 3 Beaufort Mk VIII A9-212, S-QH

A9-212 served with No. 100 Squadron for six months from February 1943. It then served with No. 7 Squadron in 1944 and No. 8 Communication Unit the following year. It ran into a culvert at Higgins Field on its way to storage in January 1946 and was converted to components.

Profile 4 Beaufort Mk VIII A9-321, T-QH, *Tom*

This aircraft served with No. 100 Squadron from July 1943 until August 1944 when it underwent repairs to its wiring. It then served with No. 32 Squadron from January 1945 until it was stored at Wagga Wagga in September 1945. It was struck off charge in August 1949.

Profile 5 Beaufort Mk VIII A9-432, V-QH, *Hallelujah Brothers - WE ARE SAVED*

A9-432 joined No. 100 Squadron in October 1943 and served until August 1944. Following an engine change, it was being flown down to Laverton with seven persons on board when it crashed near Bairnsdale in December 1944. It was destroyed and all seven were killed.

This Beaufort carries the RAF serial T9545 (it later became A9-6) and has an interesting early history. It was the first Beaufort from the Mascot Assembly Plant and was one of the six Beauforts flown to Singapore in early December 1941 before returning to Australia just a few weeks later. It joined No. 100 Squadron, RAAF, when the unit was formed in February 1942 and is seen at Richmond circa April 1942. Note that the yellow surround of the fuselage roundel has been over-painted.

For the remainder of 1943 No. 100 Squadron carried out regular reconnaissance flights and anti-submarine patrols during which submarines were occasionally attacked. These routine tasks were interspersed with bombing missions against land and sea targets, and at times the unit was exceptionally busy. Indeed, a world record was claimed for a squadron on operational service for a 30-day period when 438 sorties and 1,615 flying hours were flown from 20 July to 19 August 1943. Attacks on enemy shipping were thought to have destroyed one submarine, damaged another and possibly damaged a cruiser. In September, attacks were carried out on Gasmata, and five aircraft and crews were lost over a two day period. An additional seven aircraft were lost due to enemy action or on operational flights over the course of 1943.

In November No. 100 Squadron moved to Goodenough Island and commenced attacks on airstrips around Rabaul and shipping in Rabaul's harbour, often with Beauforts from Nos. 6 and 8 Squadrons. During one mission, Japanese aircraft followed the Beauforts back to Goodenough Island and one Beaufort was destroyed in the subsequent attack. In 1944 large scale attacks were carried out on New Britain by both day and night and again in conjunction with Nos. 6 and 8 Squadrons. These were interspersed with anti-submarine patrols, convoy escort missions and reconnaissance flights.

Starting in May 1944 aircraft began operating from Nadzab and then in June the squadron moved to Tadji on the north coast of New Guinea. Targets now were generally limited to those on and around New Guinea with the monotony varied by an attack on an enemy submarine at the end of July. At this time night harassing flights over enemy troop concentrations commenced and later in the year there was a demonstration of the dropping of supplies to Allied troops. During 1944 the squadron lost ten aircraft: including one shot down and eight lost in take-off and landing accidents.

Operations continued into 1945. Tactical reconnaissance missions, particularly over the Sepik River area, and supply dropping commenced in March. No. 100 Squadron provided a maximum air support effort in May to support the landing at Wewak. The squadron then concentrated on close support missions. To engender close cooperation with ground troops, some aircrews were attached to front line army units to gain a better understanding of the problems faced. This also enabled army officers to better understand the need to select the most appropriate run in for aircraft to obtain a clear sight of the target.

In June a detachment was sent to Biak Island to search for an enemy submarine reported in the area, but no sightings were made and crews returned after a couple of days. Upon the cessation of hostilities, No. 100 Squadron dropped leaflets over enemy occupied territory and continued with reconnaissance missions. During 1945 the squadron lost twelve aircraft including four in two mid-air collisions and another two due to the premature detonation of faulty bombs.

Starting in October 1945 some aircraft were flown back to Australia for storage and the unit started to reduce in size. However, the squadron remained busy with courier and ferry flights. At the end of January 1946, the squadron moved to Finschhafen and remained at that location for the next eight months. No. 100 Squadron disbanded on 23 August 1946, and its remaining Beauforts were destroyed in situ as it was deemed uneconomic to return them to Australia for storage.

Profile 6 **Beaufort Mk VIII A9-473, B-QH, *FIGHTIN BBs***

Having previously served with No. 1 Squadron from December 1943 until August 1944, this Beaufort joined No. 100 Squadron in November. It remained with the unit until the cessation of hostilities in August 1945. A short time later it was stored at Wagga Wagga prior to being struck off charge in August 1949.

Profile 7 **Beaufort Mk VIII A9-488, L-QH, *Let It Go Hughie!***

A9-488 served with No. 100 Squadron from January until August 1944. In January 1945 it was transferred to No. 30 Squadron (see Profile 52 on page 81). Like so many other Beauforts it was struck off charge in August 1949.

Beaufort Mk VIII A9-418 served with No. 100 Squadron from November 1943 until May 1944. It is shown here in Foliage Green and Earth Brown camouflage over Sky Blue.

Beaufort A9-142 seen at Ross River, Queensland, in 1942 in what appears to be Foliage Green and Earth Brown camouflage over Sky Blue. This bomber served with No. 7 Squadron from November 1942 for only three weeks before it ran off the end of the runway at Wards strip, Port Moresby. After lengthy repairs the bomber returned to service with No. 5 Operational Training Unit but was soon written off after an accident.

This photo shows the bomb mission markings of No. 7 Squadron Beauforts. Bomb symbols have been replaced with milk bottle symbols, based on the participant's view that the strikes they were undertaking were as routine as a "milk run".

CHAPTER 5
No. 7 Squadron

No. 7 Squadron was formed on 27 May 1940 at Laverton, Victoria, as a general reconnaissance squadron with Lockheed Hudson aircraft but these were soon withdrawn to equip other units. The squadron then remained without aircraft and existed in nucleus form only, attached to No. 2 Squadron, until January 1942. The following month the squadron again received Hudsons and commenced coastal patrols. In May the squadron moved to Bairnsdale, Victoria, and its aircraft were again withdrawn.

In August 1942, the squadron moved to Nowra, NSW, where aircrews joined it from No. 1 Operational Training Unit and Beaufort aircraft began to be received. The following month seventeen crews commenced torpedo training, but this was cut short to allow the squadron to move to Townsville. From that location No. 7 Squadron commenced coastal patrols, which were considered to be a higher priority than torpedo attack operations. The squadron gradually moved to Ross River airstrip outside of Townsville and this was completed by the end of November. Following a number of short-term commanding officers, Wing Commander Keith Parsons was appointed as commanding officer in November and would serve until early 1944.

Convoy escorts and anti-submarine patrols were a regular feature of the squadron's activities with aircraft operating from south of Townsville up to the Torres Strait and throughout the Gulf of Carpentaria. December 1942 saw three aircraft lost but without any fatalities. The squadron began detachments at Horn Island and Port Moresby, and from Horn Island a crew flying an anti-submarine patrol did sight and attack a submarine. Two further aircraft were lost in accidents during January 1943 but again there were no fatalities.

Late in January 1943, a No. 7 Squadron Beaufort located the survivors of the bombed vessel HMAS *Patricia Cam* off the coast of the Northern Territory. Emergency supplies were dropped until the survivors were rescued. Early in the New Year, further aircraft were lost, but again without fatalities. Aside from routine convoy protection missions and area patrols, other tasks included escorting single engine aircraft to Port Moresby and photographic surveys, particularly around the Merauke area. Travel flights were also flown as far afield as Coomalie Creek in the Northern Territory and Goodenough Island in the Solomon Sea.

In June 1943, six No. 7 Squadron aircrews were attached to No. 100 Squadron for one month to assist in operations over New Guinea. The month also began a series of interceptions of Japanese aircraft. Over the next six months Beauforts shot down four Jake floatplanes and damaged two others. Searches for enemy submarines were a specific aspect of activity during August. Two were sighted but could not be engaged.

The squadron then started to operate through Merauke in Dutch New Guinea and in November it began an additional role, that of bombing Japanese land targets. Aircraft conducted attacks on towns and airfields along the southern coast of New Guinea. These bombing missions continued for the next nine months. During one raid in February 1944, one aircraft was lost on the way to the target, and another was damaged by anti-aircraft fire and ditched on the return flight. All crews were rescued.

In December 1943, one aircraft was dispatched to Cairns to work with the Army Chemical Warfare Section, as part of mustard gas experiments. At this time the squadron lost its first aircrew when a Beaufort crashed into the sea during an anti-submarine patrol. In March 1944, the squadron moved to Higgins Field at the top of Cape York, allowing the detachments at Horn Island, Port Moresby and Townsville to consolidate.

During May and June 1944, the squadron was advised that it would operate two aircraft types, Beauforts and Lockheed Venturas. After some Venturas had been received, the decision was rescinded, and all Beaufort aircraft were retained. At this time, a Beaufort aircraft on an anti-submarine patrol crashed into the sea and another aircraft crashed on take-off from Higgins. The crew aboard both planes were killed. Anti-submarine patrols and convoy escorts continued, together with various other missions including radar calibration flights and supply dropping. In September 1944, the squadron began to move to Tadji on the northern coast of New Guinea. This was also the first occasion on which an RAAF squadron had been moved entirely by air, with heavy equipment coming by sea almost a year later!

From Tadji No. 7 Squadron undertook bombing missions and anti-submarine patrols. Several aircraft and two crews were lost in the next three months. Tactical reconnaissance missions commenced in March 1945 and specific naval cooperation missions the following month. The landings at Wewak in May began a period of intense operations in support of the army. These, and armed reconnaissance flights over the Sepik River area, continued until the cessation of hostilities. Squadron operations were assisted on several occasions by a local native who flew with the squadron and identified enemy gun emplacements, bivouac areas and hideouts which were subsequently attacked.

Thereafter, surrender leaflets were dropped on coastal and inland areas with No. 7 Squadron participating with other units at Tadji in the Beaufort courier service connecting Lae, Madang and Wewak. In October 1945 the squadron started to relinquish its aircraft back to Australia and wound down its activities until it disbanded at Tadji on 19 December.

A9-595 served with No. 7 Squadron from July 1944 until the cessation of hostilities. It is seen under repair at Tadji after a parking accident in July 1945.

Beaufort A9-145 ("Z-KT") joined No. 7 Squadron in January 1943. It is shown six months later when it crashed at Ross River due to engine failure on take-off. It appears to have Foliage Green and Earth Brown camouflage over Sky Blue.

This photo is interesting as it shows the two different Beaufort camouflage schemes. A9-600 ("KT-L") has disruptive Foliage Green and Earth Brown over Sky Blue while A9-647 ("KT-M") is in overall Foliage Green. Both Beauforts only joined No. 7 Squadron in 1945.

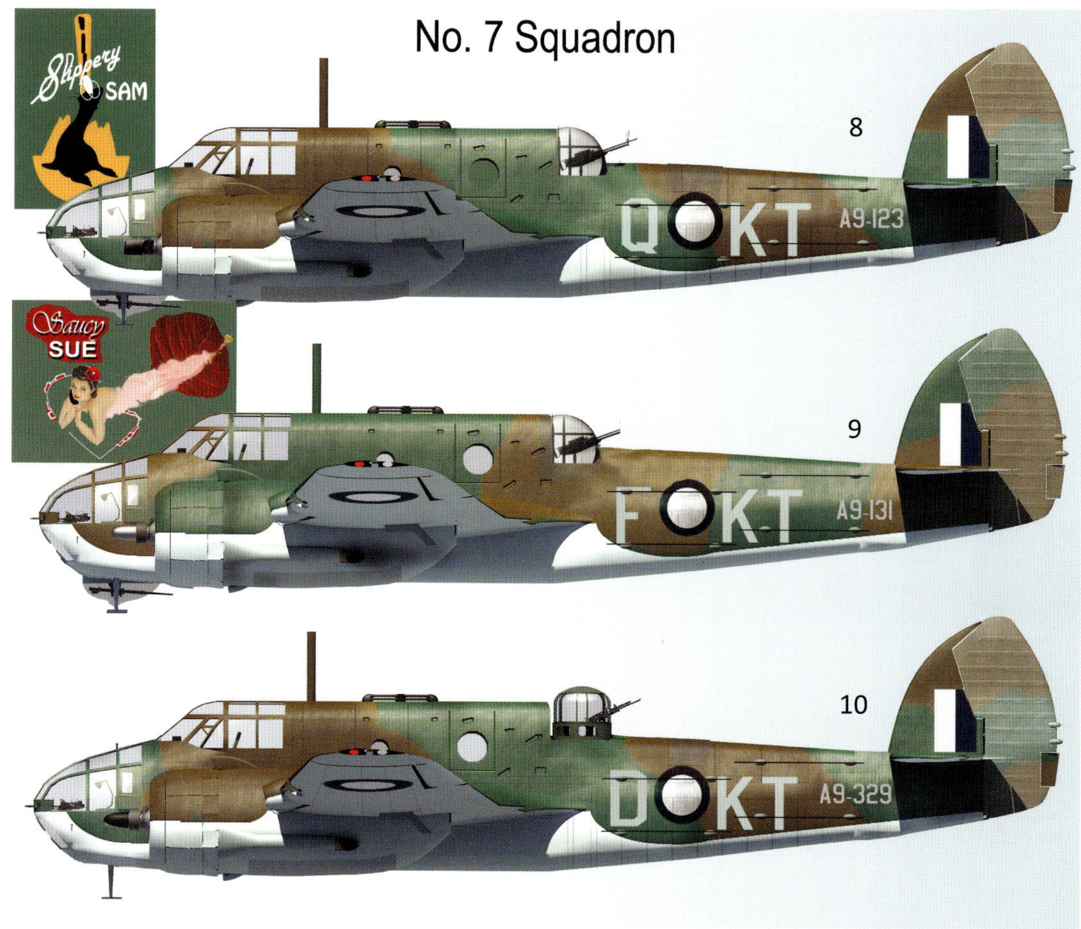

Profile 8 **Beaufort Mk VII A9-123, Q-KT, *Slippery Sam***

This bomber served with No. 7 Squadron from October 1942 until June 1943. It subsequently served with Nos. 1 and 5 Operational Training Units until it entered storage at Tocumwal in January 1946. It was struck off charge the following month.

Profile 9 **Beaufort Mk VII A9-131, F-KT, *Saucy Sue***

A9-131 joined No. 7 Squadron in October 1942 and served until June 1943. Two months later it was transferred to No. 1 Operational Training Unit, and during the last year of the war it also served with No. 9 Communication Unit. In September 1945 it was stored at Wagga Wagga prior to being struck off charge in August 1949.

Profile 10 **Beaufort Mk VIII A9-329, D-KT**

This Beaufort joined No. 7 Squadron in July 1943 and served until January 1945. It then joined No. 15 Squadron for a brief period from June 1945 until it was stored at Wagga Wagga in October 1945. Almost four years later it was struck off charge in August 1949.

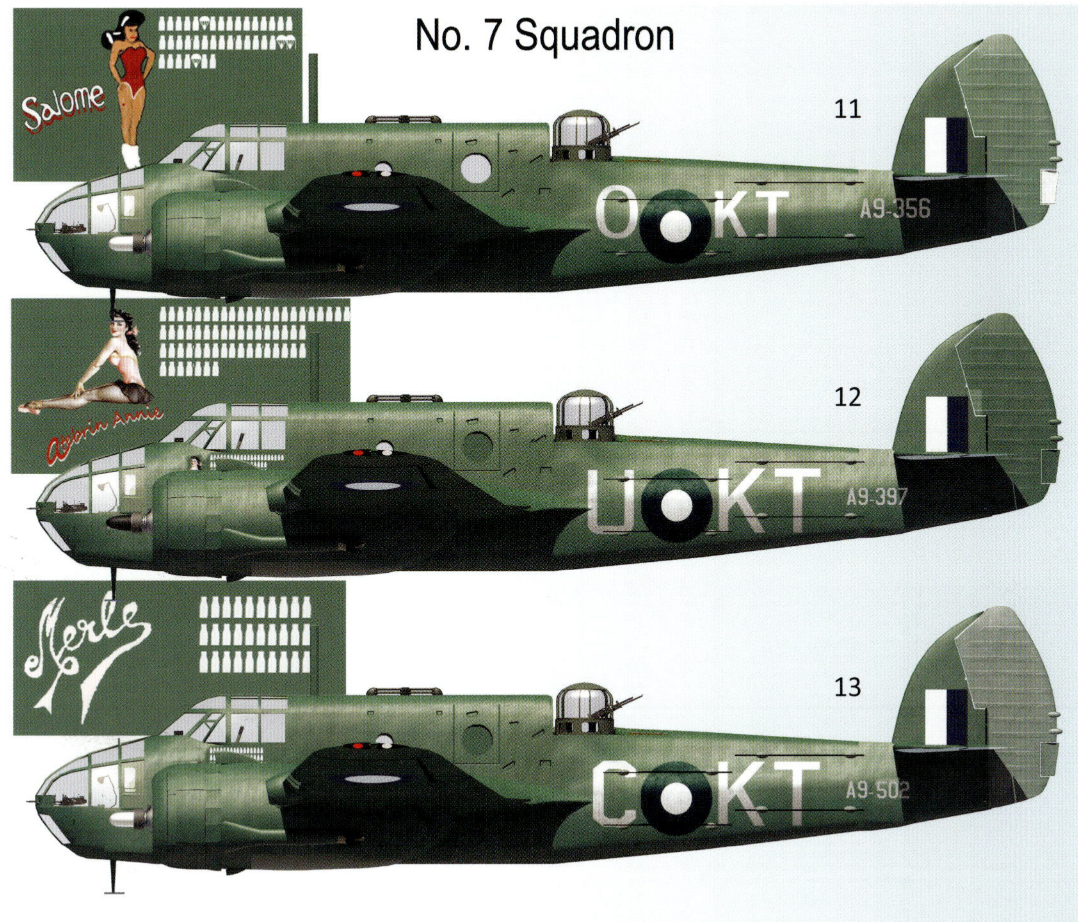

Profile 11 Beaufort Mk VIII A9-356, O-KT, *Salome*

This Beaufort originally served with No. 5 Operational Training Unit where it experienced a crash landing in August 1943. Subsequent repairs took almost a year, after which it joined No. 7 Squadron in July 1944. It remained with the squadron until June 1945. Five months later it was stored at Wagga Wagga where it remained until it was eventually disposed of in August 1949.

Profile 12 Beaufort Mk VIII A9-397, U-KT, *Atebrin Annie*

This aircraft joined No. 7 Squadron in October 1943. It remained with the unit until May 1945 when it was scheduled for inspection and repairs. After the end of the war, it was stored at Tocumwal in January 1946 and was then converted to components the following month.

Profile 13 Beaufort Mk VIII A9-502, C-KT, *Merle*

A9-502 served with No. 1 Squadron for almost a year from December 1943. During this period, it was damaged during a landing accident at Millingimbi. After repairs it eventually joined No. 7 Squadron in April 1945 and saw several months of service. In November 1945 it was stored at Tocumwal where it remained for over seven years before it was struck off charge in February 1953.

Beaufort A9-237 ("U-PN") in a typically hot and dry Western Australian setting. This aircraft served with No. 14 Squadron from March 1943 until November of that year when it crashed on landing at Onslow and was converted to components. The camouflage of Foliage Green and Earth Brown over Sky Blue is shown well here.

Beaufort A9-342 was received by No. 14 Squadron in June 1943 and was later coded "PN-F". It spent its service life with the squadron before entering storage at Wagga Wagga in September 1945. This aircraft has its under-defence gun in place - the last production Beaufort to incorporate this feature being A9-350.

Beaufort A9-539 was another that was with No. 14 Squadron for its entire service life. It joined the squadron in February 1944 and was later coded "PN-P". Like other squadron aircraft, it was flown to Wagga Wagga for storage in September 1945.

CHAPTER 6
No. 14 Squadron

No. 14 Squadron was formed at Pearce, Western Australia, on 6 February 1939 equipped initially with Avro Ansons as a general reconnaissance squadron. In 1940 these aircraft were replaced by Lockheed Hudsons and by the end of that year the squadron was operating both Hudsons and Wirraways. During December 1942, the Hudsons were gradually replaced by Beauforts, manned with crews who were already trained on those aircraft so that operations could commence almost immediately. At this time, the squadron was commanded by Wing Commander Geoff Nicholl. Convoy escort missions, reconnaissances and anti-submarine patrols were flown throughout the coastal areas of Western Australia.

The new year of 1943 did not begin well with one Beaufort crashing soon after take-off from Pearce with the loss of the crew. During January the last of the Hudson aircraft were transferred to No. 32 Squadron but there were few flights due to most Beaufort aircraft being unserviceable through a lack of spare parts. Once serviceability increased, routine patrols continued and these were interspersed with more interesting activities.

One aircraft was examining all the operational bases along the coastline up to Derby and was accompanied by a fisheries officer from the Council of Scientific and Industrial Research who was observing marine life along the coast. Another aircraft was involved on photographic flights to survey dispersal airstrips in the Perth area and a further aircraft was used to escort Kittyhawk fighters from Exmouth Gulf back to Horn Island in Queensland. Searchlight cooperation exercises were held with army units around Perth and aircraft carried out mock attacks on anti-aircraft and coastal defences. Inspection flights were also undertaken to ascertain the effectiveness of camouflaged installations throughout southwest Western Australia. Meanwhile crews honed their low-level attack tactics by bombing and strafing a target towed behind a destroyer.

By the middle of 1943, regular patrols were augmented by searches for missing vessels and two possible submarine sightings. By this time, the main patrols conducted to seaward by the squadron were from Rottnest Island off Perth and around the coast as far as Albany. The squadron now entered a period of numerous aircraft accidents, with five aircraft lost within six months. Aside from take-off and landing accidents, one aircraft went missing on a patrol and another crashed into the sea during formation flying practice due to a trim tab fault. Some of these accidents resulted in a heavy loss of life, including that of Wing Commander Charles Learmonth who had taken over command of the squadron in late 1943.

The mundane work of coastal patrols continued during 1944 and was only punctuated by the occasional search for missing aircraft, overdue shipping or reports of enemy submarines. Christmas 1944 saw one aircraft involved in travel flights to operational bases in northern Western Australia delivering Christmas poultry.

In January 1945, an extensive fire at the Fremantle docks necessitated the mooring of shipping offshore and No. 14 Squadron aircraft were ordered to protect these vessels with continuous anti-submarine patrols until the ships could return to harbour. Also in early 1945, Beauforts escorted a naval force proceeding eastward from overseas with the incoming Governor-General, the Duke of Gloucester, on board.

Soon afterwards a vessel was torpedoed by German U-Boat *U-862* with Beauforts extensively involved in searching for the submarine and for survivors. Patrols were given an added impetus by this sinking, but no submarine was sighted and routine convoy escorts and surveillance flights continued until hostilities ceased. However, squadron aircraft were used to transport fresh food to naval forces at Onslow and Learmonth and incendiaries were flown to Learmonth for use by No. 25 Squadron Liberators. One aircraft even flew on a travel flight from Pearce to Melbourne, before proceeding up the east coast and on to Morotai. It eventually returned to base two weeks later.

In September 1945 the need for No. 14 Squadron had evaporated and its thirteen Beauforts were allotted to Wagga Wagga for storage. The last Beaufort flew out at the end of October and the squadron was disbanded at Pearce on 10 December 1945.

Profile 14 Beaufort Mk VA A9-180, H-PN

Beaufort A9-180 served with No. 14 Squadron from December 1942 until September 1943. In June 1943 it forced landed near Meekatharra, Western Australia, following an engine failure. After briefly returning to the squadron, it was transferred to No. 1 Operational Training Unit where it served until it was placed in storage at Wagga Wagga in October 1945. It was struck off charge in August 1949.

Profile 15 Beaufort Mk VIII A9-275, X-PN

This bomber joined No. 14 Squadron in May 1943 and remained with the unit for over two years until July 1945. During this service it suffered a damaged tailplane in a heavy landing in June 1944. Following the end of the war it was stored at Wagga Wagga in November 1945 until it was struck off charge in August 1949.

Profile 16 Beaufort Mk VIII A9-349, O-PN

A9-349 was received by No. 14 Squadron in June 1943. It was repainted in overall Foliage Green during a major service starting in July 1944, as was standard practice for Beauforts after May of that year. It was stored at Wagga Wagga in October 1945 and struck off charge in August 1949.

Profile 17 Beaufort Mk VIII A9-358, Y-PN

This Beaufort joined No. 14 Squadron in October 1943 and a week later forced landed at Pearce due to engine failure. It eventually returned to the squadron a year later and after the war it was stored at Wagga Wagga. Like so many others it was eventually struck off charge in August 1949.

No. 14 Squadron

14

15

16

17

A9-333 joined No. 14 Squadron in August 1943 but belly landed following engine failure on take-off at Pearce in May 1944. It was not repaired until August 1945 when it rejoined the squadron for just two months before it was flown into storage at Wagga Wagga.

Beaufort A9-390 ("B-FX") was received by No. 6 Squadron in August 1943, but its service was brief. That same month the aircraft belly landed near Hollandia following engine and undercarriage failure and was written off. The camouflage scheme appears to be Foliage Green and Earth Brown over Night Black.

Beaufort A9-531 ("FX-Z") see over New Guinea during service with No. 6 Squadron in 1944. It has Foliage Green and Earth Brown camouflage over Sky Blue undersurfaces. While landing on Goodenough Island in December 1944 the bomber crashed due to a blown tyre and was converted to components.

CHAPTER 7
No. 6 Squadron

No. 6 Squadron was formed on 1 January 1939 following a change in the squadron number from No. 4. The unit was based at Richmond, NSW, and was equipped initially with Avro Ansons before receipt of Hudsons. In August 1942, the squadron moved north and operated against Japanese forces from several bases in north Queensland and New Guinea.

In August 1943, while based at Milne Bay, No. 6 Squadron began to re-equip with Beauforts. The following month, Beauforts began operational flights and within a few weeks all of the Hudsons had been transferred out and Wing Commander Colin Hannah was appointed as commanding officer. Routine tasks included convoy escort, area reconnaissance and bombing attacks. The bombing missions were often carried out against enemy shipping, sometimes in company with Nos. 8 and 100 Squadrons.

In October 1943 No. 6 Squadron moved from Milne Bay to Goodenough Island. Over the next six months missions were flown against land targets and shipping in Rabaul's harbour. There were also searches for submarines and missing aircrew, attacks on enemy destroyers and barges and message dropping flights. These missions were not without losses. Three Beauforts crashed into the sea, while several others were lost in non-fatal accidents.

In 1944 there was a lull in activity during which time training increased which included naval cooperation exercises, formation flying, fighter cooperation and bombing practice. Numerous travel flights were also initiated as far afield as Darwin. During the last half of 1944, convoy escort missions increased, together with a few bombing missions that included occasional attacks on airfields around Rabaul. At the end of the year, the squadron started to move from Goodenough Island to Dobodura. The move was completed by the end of January 1945.

Once established at its new location, bombing and strafing attacks commenced in the Sepik River area and on enemy occupied villages in New Guinea and on New Britain. This was supplemented by army support missions and night supply drops although a further three aircraft were lost. At the end of April 1945, a flight of aircraft was detached to Tadji to assist other squadrons with supporting the Australian landings at Wewak. The subsequent army support missions continued during May and a No. 6 Squadron Beaufort was the first Allied aircraft to land on newly captured Wewak airstrip.

For the remainder of the war, the squadron was not employed to its full extent. Supply drops in New Britain and photo reconnaissance missions were supplemented with a variety of training flights. Starting in July 1945, several bombing missions were carried out targeting Manokwari in Dutch New Guinea.

Following the Japanese surrender, No. 6 Squadron was one of the first overseas units to return to the Australian mainland. It arrived at Kingaroy in October 1945 where it disbanded.

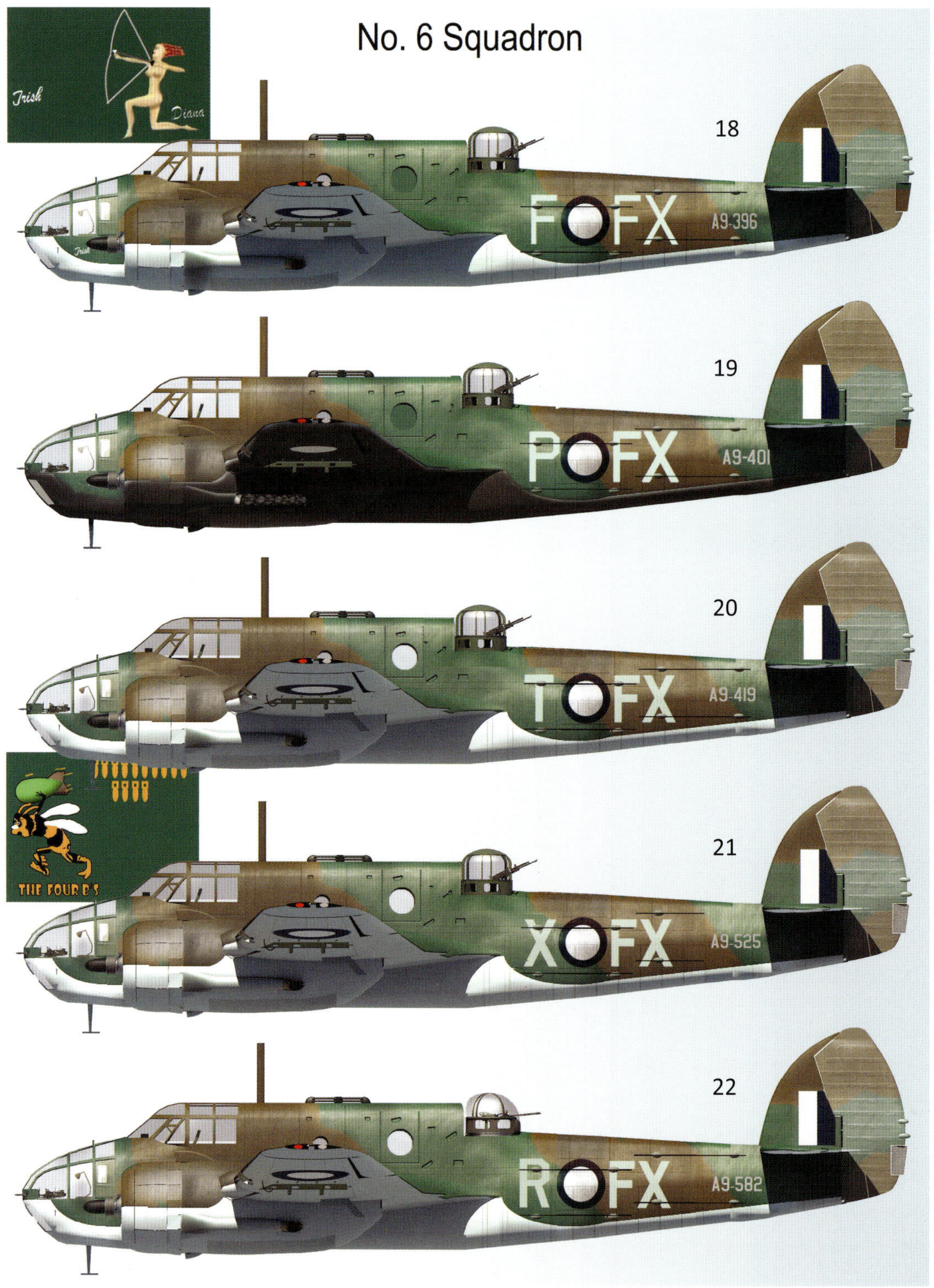
No. 6 Squadron
Irish
Diana
18
F FX
A9-396
19
P FX
A9-401
20
T FX
A9-419
THE FOUR B'S
21
X FX
A9-525
22
R FX
A9-582

Profile 18 Beaufort Mk VIII A9-396, F-FX, *Trish Diana*

A9-396 served with No. 6 Squadron from September 1943 until August 1944. It was then operated by No. 3 Communication Unit from April until September 1945 before it was flown to Wagga Wagga where it was placed in storage. It was struck off charge in August 1949.

Profile 19 Beaufort Mk VIII A9-401, P-FX

No. 6 Squadron received this aircraft in September 1943, and it served with the unit for over a year until October 1944. In July 1945 it joined No. 93 Squadron and the following month it crashed on take-off from Labuan, Borneo, and was converted to components.

Profile 20 Beaufort Mk VIII A9-419, T-FX

This Beaufort served with No. 6 Squadron from September 1943 until December 1944. In January 1945 it was being ferried to Mascot when it force landed near Taralga, NSW, and was converted to components.

Profile 21 Beaufort Mk VIII A9-525, X-FX, *THE FOUR B's*

This aircraft joined No. 6 Squadron in January 1944 but crashed on landing at Goodenough Island that December. The damage was sufficient for the aircraft to be written off and converted to components.

Profile 22 Beaufort Mk VIII A9-582, R-FX

A9-582 served less than five months with No. 6 Squadron from May 1944. It swung on landing at Milne Bay and was ground-looped to avoid hitting other aircraft that October and was converted to components.

Beaufort A9-351 joined No. 6 Squadron late in its career, in December 1944. It is shown here in 1945 at Jacquinot Bay in overall Foliage Green. Note that on the bomb rack under the wing is a storepedo, a container used to drop supplies to ground forces.

A fine aerial photo of Beaufort A9-551 ("UV-S") off the New Guinea coast in 1944. This bomber joined No. 8 Squadron in June 1944 and served until January 1945 when it was scheduled for a major service.

Beaufort A9-238 ("J-UV") seen off Goodenough Island during its service with No. 8 Squadron which commenced in April 1943. In May 1944 an engine caught fire whilst the bomber was undergoing an inspection at Nadzab and it was written off.

CHAPTER 8
No. 8 Squadron

No. 8 Squadron was formed at Canberra on 11 September 1939 and was initially equipped with Douglas DC-3 aircraft. Coastal surveillance flights were conducted almost immediately. In May 1940 the first Lockheed Hudson aircraft was received and by August that year, now fully equipped with Hudsons, the squadron moved to Singapore. Following the Japanese entry into the war, the squadron was keenly engaged in operations until it moved to the Netherlands East Indies in January 1942. The remaining personnel were soon returned to Australia as the squadron had suffered such losses it was no longer effective.

No. 8 Squadron was not reformed until March 1943 at Canberra as a general reconnaissance/ torpedo squadron and its initial commanding officer was Wing Commander Geoff Nicholl. It started re-equipping with Beaufort aircraft and training operations commenced. This continued until May when the squadron moved to Bohle River, Townsville. However, the training period was not without losses, with one aircraft crashing at Canberra and a further aircraft crashing into the sea but with all of the crew rescued safely.

During June and July 1943 No. 8 Squadron conducted coastal patrols but during this period a further two aircraft were lost: one crashed into the sea off Townsville, and another crashed at Bohle River. During July through to September the squadron moved north to Goodenough Island, but again an aircraft was lost when it failed to return after bombing a Japanese camp and supply dump at Cape Hoskins, New Britain. In October one Beaufort was destroyed during an enemy air raid. Regular New Britain bombing missions continued and following an attack on Buka airfield, one aircraft crashed into the sea. The squadron now reverted to torpedo attacks on shipping near Buka Passage and on shipping in Rabaul's harbour. However, torpedo malfunctions were a continuing problem.

Torpedo attacks, searches for submarines and supply dump bombing missions were standard operations for the remainder of 1943. A torpedo strike on shipping off the east coast of New Ireland resulted in one ship probably sunk but at the cost of one aircraft lost. Another aircraft was lost on a bombing mission, but fortunately the crew was rescued. During November and December Rabaul was a target for bombing attacks and also a torpedo attack on Simpson Harbour shipping with another aircraft lost. A further torpedo attack on warships resulted in one probably sunk and hits scored on two motor vessels in Blanche Bay although one Beaufort did not return. After another bombing attack on Christmas Day, an aircraft returned to Goodenough Island. However, a bomb that failed to release fell off when the aircraft landed and exploded, destroying the aircraft and killing the crew.

Early 1944 saw large scale squadron attacks on New Britain involving up to twenty Beauforts. The majority of these bombing missions were at night with the aircraft now having their undersurfaces painted black. Further aircraft losses occurred during January, although torrential rain mitigated against further operations for a period as the Goodenough Island airstrip and nearby roads were badly damaged.

March 1944 saw close air support provided for US Marines attacking Talasea in New Britain. Two months later eleven Beauforts flew from Goodenough Island to Nadzab to establish an air echelon to support operations in the Wewak area. Attacks commenced almost immediately although three aircraft were lost. In June No. 8 Squadron Beauforts moved from Nadzab further west along the coast to Tadji. Operations commenced in support of ground advances, although it was to be a further three months before the squadron ground personnel joined them. From this time until the end of the year, a further three aircraft were lost. In an attack on the Wewak area, three aircraft were damaged due to bomb blasts from their own bombs during dive-bombing attacks.

From Tadji No. 8 Squadron began attacking Japanese held villages and airstrips, as well as escorting small convoys with anti-submarine sweeps along the coast. However, a detachment of aircraft did return to Goodenough Island to operate from there against Rabaul for a temporary period and to participate in supply dropping over New Britain.

Commencing in early 1945, operations included attacking barges, bombing enemy villages, searches for missing aircraft, supply and leaflet dropping missions and anti-submarine sweeps, during which another Beaufort was lost. Armed tactical reconnaissances of the Sepik River area in search of targets of opportunity became a regular occurrence until the cessation of hostilities. As with other Beaufort squadrons based at Tadji, the squadron commenced operations in very close support of the army in the Wewak area with targets indicated by mortar rounds, sometimes within only a hundred metres of friendly positions.

A photographic mosaic of the Wewak area was completed during May with the main Japanese headquarters being bombed and strafed from which a reasonable amount of anti-aircraft fire was encountered. One member of the squadron was drowned in the surf adjacent to their campsite in July, despite a surf lifesaving club having been formed in the area a few months before.

Following the cessation of hostilities, No. 8 Squadron crews participated in searches for missing aircraft, leaflet dropping missions and reconnaissance flights. The squadron started to slowly contract in size with some aircraft returning to Australia and others continuing with a photographic survey of Sepik River area. Four Beauforts were damaged in accidents during October and November before all remaining aircraft were ferried to Tocumwal, NSW, for storage. After the posting of remaining air and ground crew to other units, No. 8 Squadron disbanded on 19 January 1946.

Beaufort A9-316 ("UV-N") seen at Tadji, New Guinea, during its service with No. 8 Squadron from May 1943 until October 1944. It was later modified as a Mk IX "Beaufreighter" in July 1945 before it was destroyed in a crash landing at Piva, Bougainville, in November of that year.

Beaufort A9-640 seen at Tadji, New Guinea, in the overall Foliage Green scheme. This aircraft served with No. 8 Squadron from September 1944 until it was flown into storage at Tocumwal in January 1946.

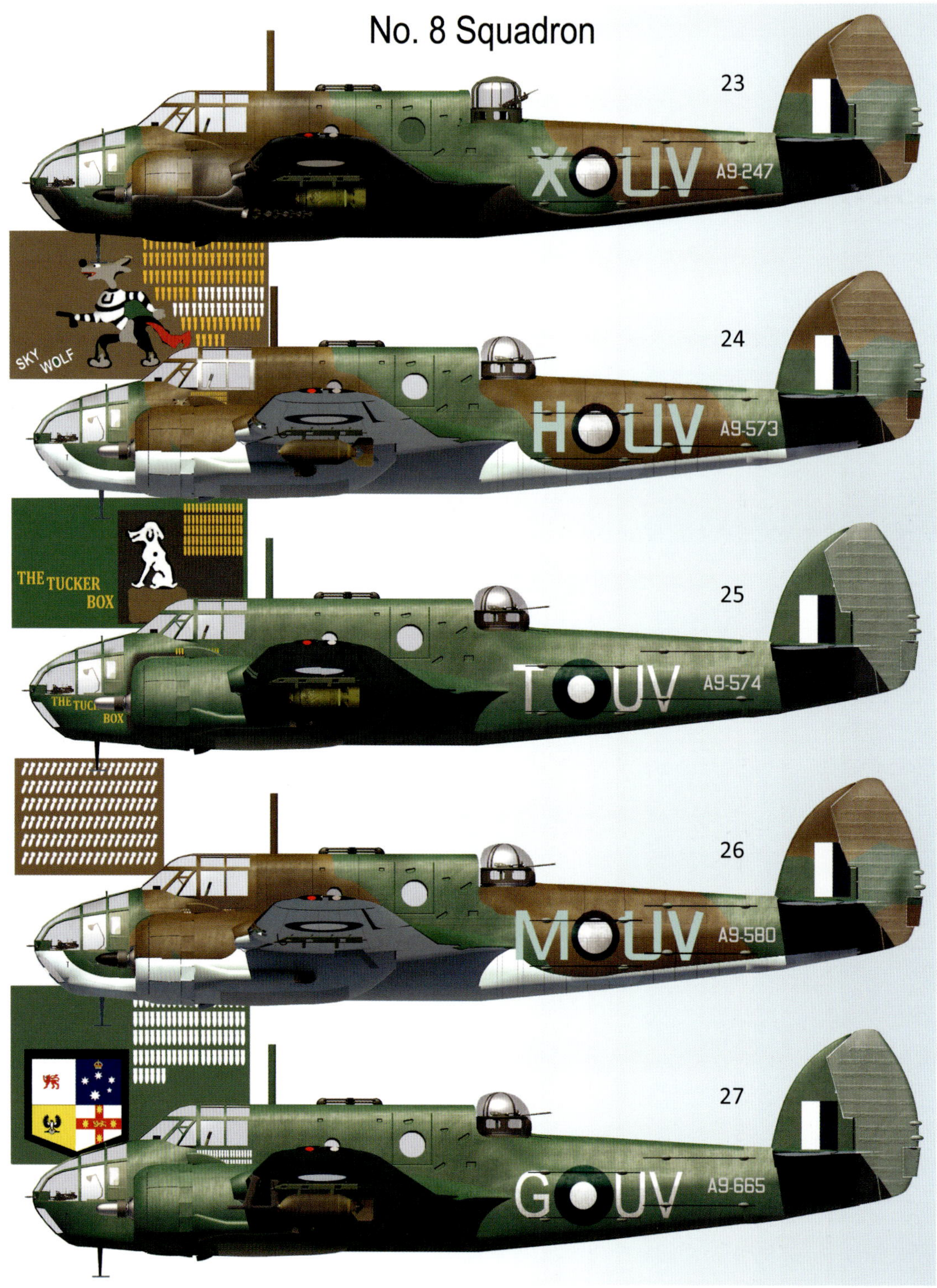
No. 8 Squadron
23
XOUV
A9-247
SKY WOLF
24
HOUV
A9-573
THE TUCKER BOX
25
TOUV
A9-574
26
MOUV
A9-580
27
GOUV
A9-665

Profile 23 Beaufort Mk VIII A9-247, X-UV

A9-247 joined No. 8 Squadron in April 1943. On 8 November of that year it was one of three Beauforts that attacked shipping in Simpson Harbour, Rabaul, with torpedoes. The aircraft and its crew were lost.

Profile 24 Beaufort Mk VIII A9-573, H-UV, *SKY WOLF*

This Beaufort served with No. 8 Squadron from July 1944 until January 1946 when it was flown into storage at Tocumwal. It was struck off charge in February 1953.

Profile 25 Beaufort Mk VIII A9-574, T-UV, *THE TUCKER BOX*

This aircraft joined No. 8 Squadron in January 1945 and served until after the end of the war. It has previously been damaged by enemy action during service with No. 100 Squadron. After several years in storage at Tocumwal it was finally disposed of in February 1953.

Profile 26 Beaufort Mk VIII A9-580, M-UV

A9-580 flew 142 sorties with No. 8 Squadron during six months of service in 1944. After the war the airframe seemed likely to become one of a modest number of surviving Beauforts when it was set aside for the Australian War Memorial. However, it was stored outdoors and steadily deteriorated before the memorial abandoned its plans to display it in 1953. Despite a brief reprieve, A9-580 fell victim to vandalism and exposure to the weather and had been sold for scrap by the early 1960s.

Profile 27 Beaufort Mk VIII A9-665, G-UV

This bomber first served with No. 8 Squadron from January 1945 until it was transferred to No. 100 a year later. It remained with that squadron until June 1946 when it was destroyed at Finschhafen, with the other squadron aircraft, as it was deemed uneconomic to return them to Australia.

A9-674 ("UV-X") joined No. 8 Squadron in May 1945. After the cessation of hostilities, the aircraft was being flown into storage when it forced landed due to engine failure near Coopernook, NSW, as shown above. It was converted to components.

This No. 13 Squadron detachment was photographed at Coffs Harbour in late 1943. The aircraft were based there to conduct anti-submarine sweeps off the east coast. The aircraft are A9-388 ("SF-K"), A9-373 ("SF-D") and A9-361 ("SF-A"). All have Foliage Green and Earth Brown camouflage over Sky Blue undersurfaces.

No. 13 Squadron Beaufort A9-380 ("SF-H") at Canberra in the latter part of 1943, shortly before it was damaged during a take-off accident at that location. It then spent just over a year in storage before it was written off.

CHAPTER 9
Eastern Australia Beaufort Squadrons

This chapter summarises two operational Beaufort units based in eastern Australia: Nos. 32 and 13 Squadrons.

No. 32 Squadron was formed on 21 February 1942 at Port Moresby equipped with Hudsons. It later relocated to Townsville before moving south to New South Wales, where it was based at Richmond and then Camden. At the last location in March 1943 the squadron began re-equipping with Beauforts. All Hudsons had been transferred out by the middle of May 1943, and during this period Wing Commander Peter Parker was the commanding officer.

The squadron soon commenced anti-submarine patrols, convoy escort missions and offensive sweeps. To assist with these activities along a vast stretch of the eastern Australian seaboard, Beauforts were forward based at locations from Mallacoota in Victoria right through to Bundaberg in Queensland. These activities continued until the end of the war and were supplemented by naval cooperation flights, land and marine photography, radar calibration flights, army cooperation exercises and dinghy dropping tests.

The squadron had a number of accidents during this time, some relatively minor and several fatal. The losses included two aircraft which collided during formation flying practice and another that flew into a mountain. There was action in June 1943 when a Beaufort crew attacked and probably sank the submarine *I-178* off the NSW coast.

Training continued when No. 32 Squadron was not engaged in operations. This included bombing and strafing, air to air gunnery exercises and occasional travel flights. Some of the travel flights involved taking VIPs to northern operational areas, including as far afield as Morotai. In early May 1944 the squadron prepared to move from Camden to Lowood, Queensland, and operations commenced from the new base later in the month. A detachment remained at Bundaberg until September before joining the remainder of the squadron at Lowood. In January 1945, a detachment was sent to Laverton, Victoria, in response to increased submarine activity over that summer but the detachment was withdrawn after only a few weeks.

No. 32 Squadron participated in the protection of the British Pacific Fleet in its movement from Sydney to join the US naval forces in the north Pacific. The tedium of patrols and convoy escorts continued until these became unnecessary following the Japanese surrender in August. The squadron was disbanded at Lowood on 26 November 1945.

No. 13 Squadron was formed at Darwin on 1 June 1940, initially with Avro Ansons but soon also with modern Hudsons which were very actively employed to the north of Australia following the declaration of war with Japan. Intense operations continued from the Northern Territory until the beginning of April 1943. The squadron was then scheduled to re-equip with new aircraft types and subsequently reformed at Canberra in July 1943 with one flight of

Beauforts and one flight of Lockheed Venturas. At this time Wing Commander Peter Parker was transferred in as the commanding officer.

Several experienced aircrews arrived during August, and No. 13 Squadron continued to operate with two types, having ten Beauforts and five Venturas on strength by the end of September. During that month, training proceeded with night, local and cross country flying. For the months of November and December 1943, a flight of Beauforts was based at Coffs Harbour and at Evans Head conducting anti-submarine patrols and offensive sweeps in response to the increase in Japanese submarine activity along the eastern Australian coast. Searches for missing aircraft were also conducted. All these operational flights were undertaken by Beauforts. During December 1943 further Venturas were received and the Beauforts were gradually assigned elsewhere, many to No. 2 Squadron which was relinquishing their Hudsons. All Beauforts had been transferred out of the unit by the end of December.

No. 13 Squadron continued to fly Venturas until the end of the war from several different bases in Australia and to the north, before disbanding at Labuan, Borneo, on 11 January 1946.

A pair of No. 32 Squadron Beauforts over the Queensland coast around December 1944. The aircraft are A9-609 ("JM-S") and A9-457 ("JM-X"). Both are in overall Foliage Green.

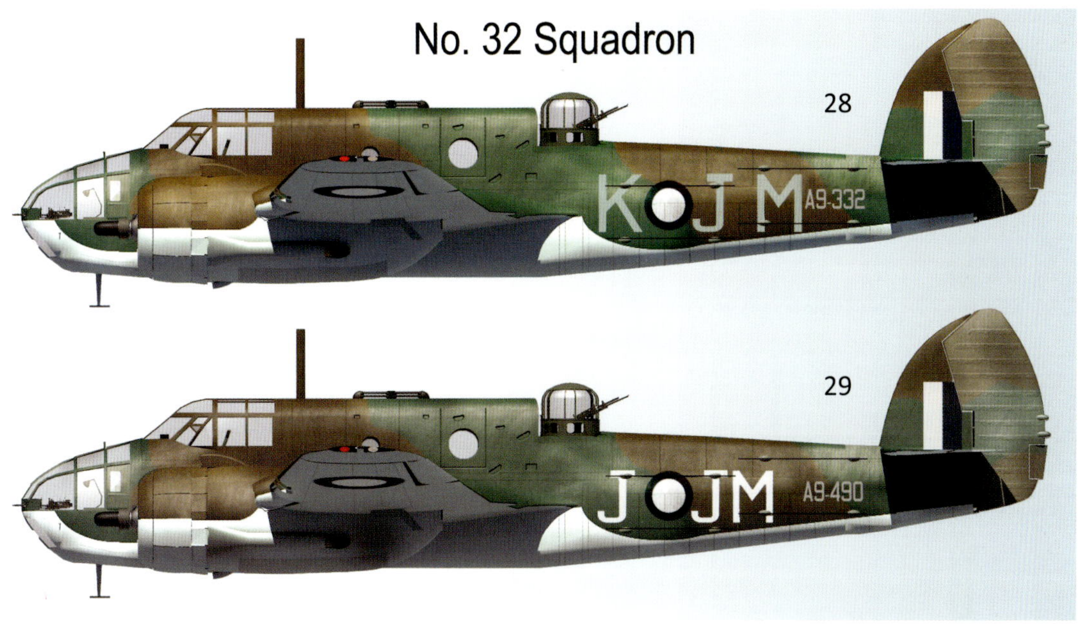

Profile 28 **Beaufort Mk VIII A9-332, K-JM**

A9-332 served with No. 32 Squadron from May 1943 until July 1945 when it was scheduled for a major service. However, it remained at Richmond until it was stored there in August 1946.

Profile 29 **Beaufort Mk VIII A9-490, J-JM**

This bomber joined No. 32 Squadron in November 1943 and continued to operate with the unit until April 1945 when it was scheduled for a major service. It was at Wagga Wagga at the cessation of hostilities and was placed into storage there in February 1946.

Beaufort A9-251 ("JM-W") which joined No. 32 Squadron in April 1943. In October of that year the aircraft ground looped on landing at Williamtown following undercarriage failure, as shown. It was damaged beyond repair and converted to components. It has Foliage Green and Earth Brown camouflage over Sky Blue.

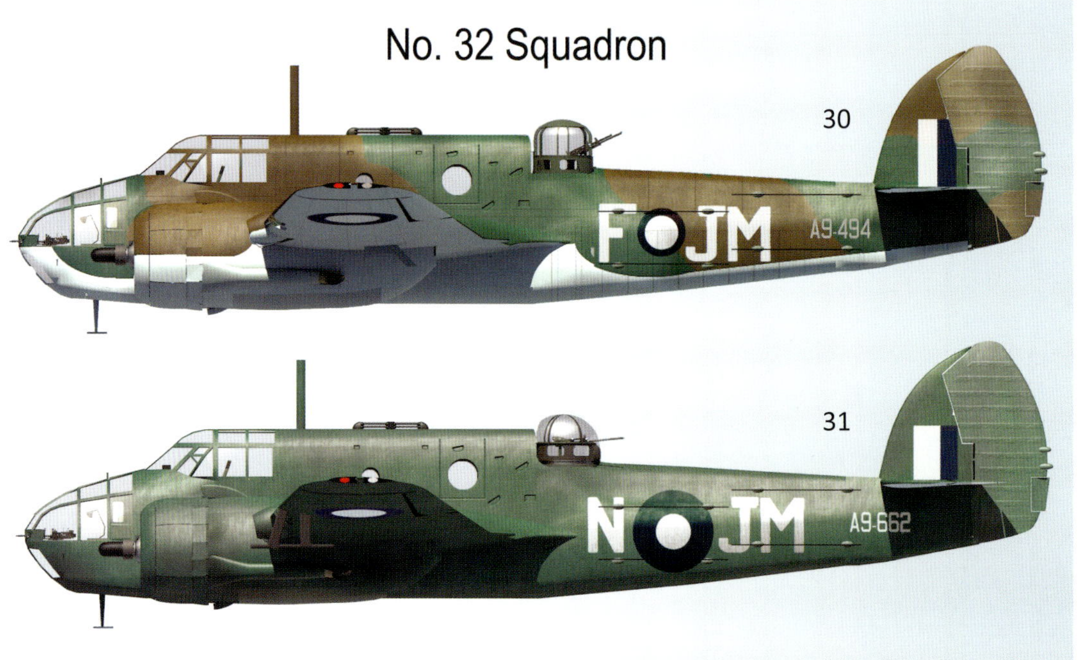

Profile 30 Beaufort Mk VIII A9-494, F-JM

This Beaufort served with No. 32 Squadron from December 1943 until it also required a major service at Wagga Wagga in March 1945. It was stored there several months later and was struck off charge in August 1949.

Profile 31 Beaufort Mk VIII A9-662, N-JM

This aircraft joined No. 32 Squadron in December 1944 and was involved in a landing accident at Lowood in May 1945. It was subsequently stored at Wagga Wagga in September 1945 and was converted into an instructional airframe in November 1947.

A line up of No. 32 Squadron Beauforts at Lowood with A9-563 in the foreground in overall Foliage Green.

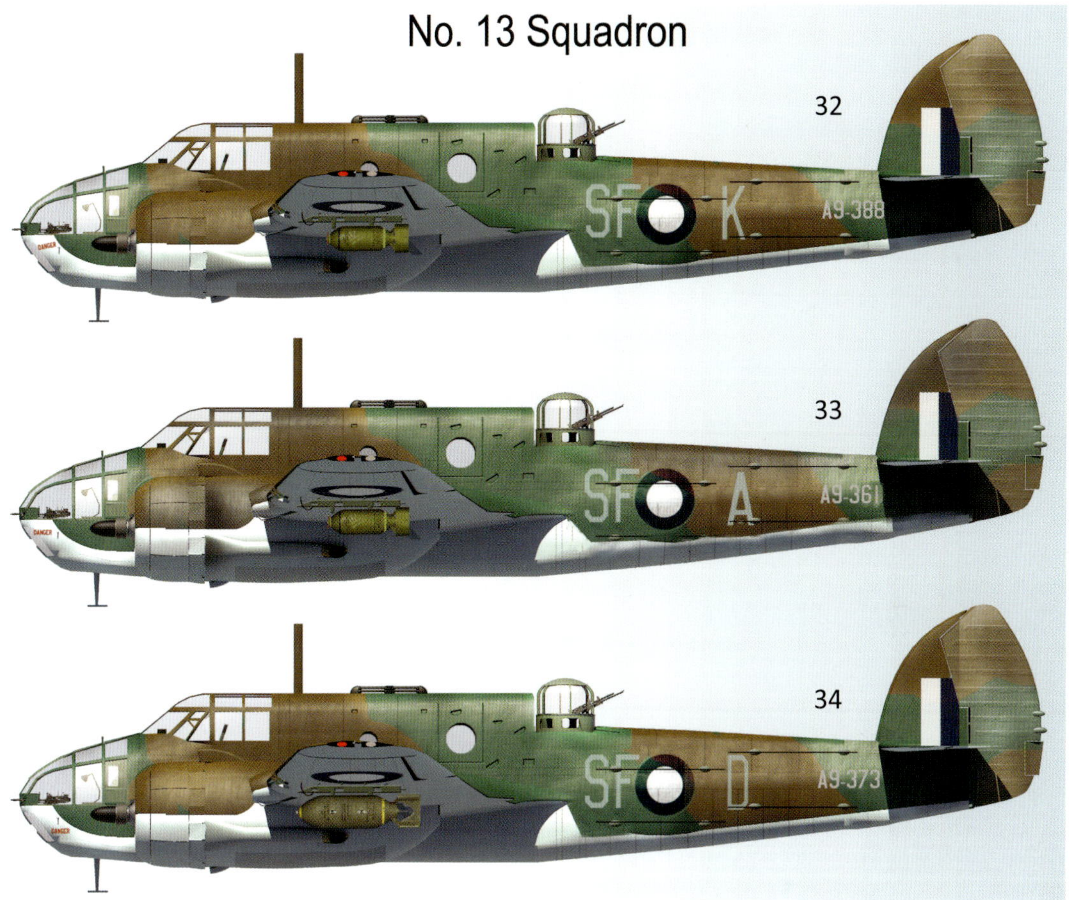

Profile 32 Beaufort Mk VIII A9-388, SF-K

This bomber commenced its service with No. 13 Squadron in August 1943 but in December of that year it was transferred to No. 2 Squadron. Six months later it joined No. 1 Squadron before moving again to No. 1 Operational Training Unit in 1945. After a few years in storage at Wagga Wagga it was struck off charge in August 1949.

Profile 33 Beaufort Mk VIII A9-361, SF-A

A9-361 served with No. 13 Squadron from August until December 1943 and was subsequently operated by Nos. 1 and 2 Squadrons the following year. It was scheduled to be reconditioned in early 1945 but was instead received by a number of maintenance units until it entered storage at Wagga Wagga. In common with many others, it was struck off charge in August 1949.

Profile 34 Beaufort Mk VIII A9-373, SF-D

This aircraft joined No. 13 Squadron in August 1943 with which it served for around five months. It was then passed between Nos. 1, 2 and 8 Squadrons, and in 1945 it was fitted with dual controls for training. In January 1946 it again moved to No. 100 Squadron, but it crash landed at Tadji that month and was converted to components.

Beaufort A9-501 ("P-NA") in what appears to be an overall Foliage Green scheme with possibly Night Black undersurfaces. This photo was likely taken not long after it left No. 1 Squadron.. The bomber had joined the unit in December 1943 and remained until the following November. Four months later in March 1945 it was written off after a crash landing at Gove.

Beaufort A9-540 ("NA-L") served with No. 1 Squadron from May 1944 until January 1945. This bomber survived the war and remained in storage at Wagga Wagga until it was struck off charge in 1953. It is seen in a Foliage Green scheme, with what appears to be Night Black undersurfaces.

CHAPTER 10

Northern Territory Beaufort Squadrons

This chapter summarises two operational Beaufort units that served in the Northern Territory: Nos. 1 and 2 Squadrons.

No. 1 Squadron was formed on 1 July 1925 at Point Cook, Victoria, and operated various aircraft types until it received its first Lockheed Hudsons in March 1940. Three months later the squadron began moving to Malaya. From December 1941 No. 1 Squadron fought the brunt of the Japanese advance. After taking heavy losses, it withdrew to the Netherlands East Indies but was effectively destroyed as a fighting force. A few Hudsons managed to escape to Australia before many squadron personnel were captured when the Japanese invaded Java.

The squadron reformed at Menangle, NSW, on 1 December 1943 under the command of Wing Commander David Campbell. That month it received its first Beauforts and a training program commenced soon after. An unfortunate early accident occurred when two Beauforts collided mid-air on a cross-country flight and were both destroyed. In February 1944 the squadron commenced its move to Gould in the Northern Territory where it was quickly operational. During 1944 the squadron's area of operations extended from Truscott in Western Australia across to Horn Island in Queensland.

The activities required of No. 1 Squadron were significant. They consisted of bombing and strafing missions against various targets in Timor and other islands to the north of Australia, sometimes at night. Photo reconnaissance flights around enemy coastal and inland areas were also undertaken, alongside the usual anti-submarine patrols, convoy escort missions and other sundry tasks.

Given the high number of hours flown the squadron suffered relatively few losses. One aircraft crashed shortly after becoming airborne on an early morning flight and another aircraft ditched into the sea with the crew being rescued shortly after. In January 1945, No. 1 Squadron ceased operations and moved to Kingaroy, Queensland, to re-equip with Mosquito aircraft. The Beauforts was ferried to numerous units across Australia. The squadron subsequently operated from Morotai and Labuan until the cessation of hostilities. It then moved back to Narromine, NSW, where it disbanded in June 1946.

No. 2 Squadron was re-formed at Laverton, Victoria, in May 1937 with a collection of three different aircraft types. It was later equipped with Avro Ansons and in July 1940 it received Hudsons. In late 1941 it moved to Darwin with detachments operating from islands in the eastern Netherlands East Indies. Following the Japanese occupation of Timor in February 1942 the squadron moved back to the Northern Territory where it continued operations with Hudsons for the next two years.

As the Hudsons became worn out, it was decided to replace them with Beauforts, a process which started in December 1943. In January 1944, Wing Commander Les Ingram was appointed commanding officer, with Beauforts arriving from No. 13 Squadron which was re-equipping with Lockheed Venturas. No. 13 Squadron Beaufort crews were transferred with their aircraft which minimised the time taken to ready crews for operations. By this time, No. 2 Squadron was based at Hughes, south of Darwin. The Beauforts undertook convoy escort missions, maritime patrols, supply dropping flights and bombing of targets in Timor. Occasionally enemy aircraft were encountered.

No. 2 Squadron did not retain its Beauforts for long. The same month that the last Hudson left, April 1944, the squadron began to re-equip with B-25 Mitchells and conversion courses commenced immediately. By early June all Beauforts had been assigned to No. 1 Squadron or to other units in the area. No. 2 Squadron continued with its Mitchells for the remainder of the war and moved from Hughes to Balikpapan, Borneo, in August 1945. It then returned to Australia in December and was disbanded at Laverton in May 1946.

Profile 35 Beaufort Mk VIII A9-560, KO-V

A9-560 joined No. 2 Squadron in April 1944 but a month later it swung on take-off from Gould with the starboard undercarriage collapsing. It does not have the distinctive yellow fuselage codes adopted by some No. 2 Squadron Beauforts, as seen in the photo below.

Beaufort A9-576 seen under camouflage netting. This bomber joined No. 2 Squadron in March 1944. The squadron codes appear to be in the distinctive No. 2 Squadron style of yellow outlined in black.

Profile 36 Beaufort Mk VIII A9-375, V-NA

This aircraft initially served with Nos. 13 and 2 Squadrons before it was taken up by No. 1 Squadron in May 1944. In February 1945 it was transferred once again to No. 82 Wing before it was placed in storage after the end of the war. It was struck off charge in August 1949.

Profile 37 Beaufort Mk VIII A9-487

A9-487 was received by No. 1 Squadron in December 1943. In January 1945 it was reconditioned, and six months later it was issued to No. 11 Communication Unit. In April 1946 it was stored at Tocumwal and the aircraft was eventually struck off charge in February 1953.

Profile 38 Beaufort Mk VIII A9-482, D-NA

This Beaufort served with No. 1 Squadron from December 1943 until January 1945. It then joined No. 100 Squadron in April 1945. In January 1946 it ran off the end of Tadji airstrip after brake failure and was converted to components.

An aerial shot of Beauforts A9-530 ("N") and A9-548 ("L"), taken soon after the pair joined No. 15 Squadron in February 1944. Both aircraft are in Foliage Green and Earth Brown camouflage over Sky Blue undersides. The individual aircraft letter codes have been applied but not the squadron "DD" codes. Note Profile 41 of A9-548 appears on Page 72. In July 1944 all No. 15 Squadron Beauforts were repainted overall Foliage Green.

A frontal view of A9-514 that served with No. 15 Squadron from February 1944 until October 1945. The motif shows a running dog over the phrase "She's Caster". In the 1940s, "She's caster" was interchangeable with the phrase "She's apples".

CHAPTER 11
No. 15 Squadron

No. 15 Squadron was formed at Camden, NSW, in January 1944 and it immediately received Beauforts. By the end of February, formation flying and fighter cooperation exercises commenced, and the squadron was soon under the command of Wing Commander Geoff Newstead, an appointment which lasted for the next fourteen months. The squadron then moved from Camden to nearby Menangle where training continued. This was marred by a fatal accident when one aircraft crashed shortly after take-off.

The squadron was sufficiently trained in April for it to begin anti-submarine patrols and convoy escort duties along the NSW coast, which continued until February of the following year. In the meantime, the squadron moved back to Camden during May 1944. Aircraft from the squadron were detailed to escort Spitfires from Nos. 548 and 549 Squadrons, RAF, from Townsville to airfields south of Darwin. In July an advanced party commenced the move by sea of the squadron to Madang, but it took two months for the party to arrive.

Meanwhile training and patrols continued. In December 1944, a Liberty ship was torpedoed near Sydney and No. 15 Squadron Beauforts were quick to locate survivors. These patrols continued throughout January 1945 and squadron activity intensified with the escort of numerous ships of the British Pacific Fleet.

Finally, in February 1945 the squadron received orders to move to Madang. One Beaufort was lost in transit off the Queensland coast, and soon after arrival at the new base another went missing over Dutch New Guinea. Routine missions included mail deliveries, reconnaissance, supply dropping and bombing strikes. Throughout May the squadron assisted in bombing missions around Wewak in support of the landing of Australian troops.

In June 1945, the squadron started its move to Middleburg Island near the extreme western end of Dutch New Guinea. The following month, bombing attacks commenced on several towns in Dutch New Guinea. Here the squadron was making its presence felt with these attacks which were interspersed with searches for missing aircraft and leaflet dropping.

After the cessation of hostilities, the squadron was notified that it would be returning to Australia. By the end of the September most of its aircraft had been flown out and the squadron moved to Kingaroy, Queensland, the following month. There the squadron was disbanded in March 1946.

No. 15 Squadron
39
S DD
A9-510
40
F DD
A9-538
41
L DD
A9-548
15
Snow Goose
42
T DD
A9-632

Profile 39 Beaufort Mk VIII A9-510, S-DD

This Beaufort served with No. 15 Squadron from February 1944 until it was stored at Wagga Wagga in October 1945. It was struck off charge in August 1949. It is relatively unusual for an operational Beaufort to have had its camouflage removed, leaving a bare metal finish.

Profile 40 Beaufort Mk VIII A9-538, F-DD

Similar to A9-510 above, this aircraft joined No. 15 Squadron in February 1944 and remained with the unit until it was flown into storage at Wagga Wagga in October 1945. It too was struck off charge in August 1949.

Profile 41 Beaufort Mk VIII A9-548, L-DD

A9-548 was also received by No. 15 Squadron in February 1944 and was operated until it was stored at Wagga Wagga in October 1945. Like so many others it was struck off charge in August 1949. A photo of this aircraft, simply coded "L", appears on page 70.

Profile 42 Beaufort Mk VIII A9-632, T-DD, *Snow Goose*

This bomber first served with No. 100 Squadron and after a minor accident it was transferred to No. 15 Squadron in June 1945. Following the cessation of hostilities, it returned to No. 100 Squadron in September 1945 until it was flown to Wagga Wagga for storage the following month. In common with many others, it was struck off charge in August 1949.

Beaufort A9-530 ("N-DD") seen at Middleburg Island in 1945. Shown in overall Foliage Green, the bomber served with No. 15 Squadron from February 1944 until it went into storage at Wagga Wagga in October 1945.

Beaufighter Mk Ic A19-2 served with No. 30 Squadron from June 1942 until February 1943 and was later coded "B". It is seen here at Wards strip in Port Moresby. Images from this period show the camouflage in a weathered condition thus indicating it may have retained its RAF Dark Green and Dark Brown camouflage, which deteriorated badly in the tropical conditions.

Beaufighter Mk Ic A19-4, later coded "D", seen at Charters Towers circa mid-1942. It appears to be in its original RAF Dark Green and Dark Brown scheme including national markings. This aircraft served with No. 30 Squadron from June until November 1942 when it was damaged by enemy action resulting in a wheels-up landing at Port Moresby.

CHAPTER 12
No. 30 Squadron

No. 30 Squadron was formed in March 1942 at Richmond, NSW, but did not start to receive its Beaufighter aircraft until early June, when Wing Commander Brian Walker was appointed as commanding officer. Conversion to the new aircraft began immediately but training was initially slow due to the limited number of aircraft available. Unserviceability resulting from these new aircraft and their engines caused further delays. To assist with pilot conversion, a Beaufort was received in July and the first reconnaissance flights to seaward commenced. In August the squadron moved from Richmond to Bohle River, Queensland, where formation flying, cross country and sea reconnaissance flights were undertaken with aircraft on standby for night operations if Japanese air raids resulted.

The following month, the squadron started to move to Port Moresby and commenced operations against coastal vessels, troop concentrations and supply dumps along the Kokoda Track and airstrips at Buna and later Lae. These operations continued for the remainder of the year but the squadron was also on the receiving end, with Port Moresby experiencing several air raids. During 1942 eight Beaufighters were lost, with most the result of enemy action.

Operations in 1943 began with low-level attack missions on Lae airstrip and Japanese camps and supply dumps, together with regular coastal sweeps searching for barges. However, the Beaufighters were subject to increasing anti-aircraft fire and air opposition from Japanese fighters.

No. 30 Squadron had a major part to play in what became known as the Battle of the Bismark Sea, an attack on a reinforcement convoy from Rabaul to Lae. The lead up to the operation saw coordinated training exercises with US aircraft targeting a wrecked ship off Port Moresby. The Beaufighters had a key role in leading a low-level attack force that decimated all of the Japanese transports. In the following days, the squadron destroyed surviving lifeboats and barges.

Further operations included attacks on Madang and Gasmata airstrips, and against enemy troop concentrations in the Mubo area. In the middle of 1943 No. 30 Squadron relocated to Goodenough Island. However, it suffered from serviceability issues but continued with convoy escorts, armed reconnaissances and barge sweeps. Towards the end of 1943, the squadron was called upon to carry out intensive operations against the enemy and shot down two aircraft. After the Japanese conducted a number of air raids on Goodenough Island the squadron moved to Kiriwina in November. During 1943 the squadron lost 30 Beaufighters, with over half due to combat.

Early in 1944 operations resumed with barge sweeps and searches for enemy vessels including submarines. One submarine was attacked with bombs as it was submerging and was possibly damaged. Enemy troop concentrations and villages were bombed and strafed and leaflet dropping commenced.

In May 1944 the squadron commenced moving to Tadji and operations started immediately with barge sweeps and attacks on land targets around Wewak. Cover was also provided for US PT boats, together with armed reconnaissance patrols. By the middle of the year, a lack of engine oil at Tadji grounded all aircraft before the squadron moved to Noemfoor Island where the same type of operations continued with attacks up to Ceram and the Celebes.

During October No. 30 Squadron aircraft were fitted with rocket projectiles and training commenced in their use. At this time the squadron was conducting operations over a wide area, from Morotai across to Middleburg Island. During November the squadron moved again, this time to Morotai. Attacks commenced on oil tanks, radar installations, buildings, watercraft, supply dumps, airstrips and ship building yards, with rockets now being used operationally. During 1944, 29 Beaufighters were lost, with fourteen of these losses due to enemy action, including two destroyed as a result of an enemy raid on Morotai.

Intruder missions from Morotai continued into 1945 with villages, barge hideouts, oil dumps, radio stations, bridges, power houses and barracks targeted. The arsenal of weapons available was further enhanced with napalm being used in addition to conventional bombs and rockets. In addition, cover for convoys was provided and courier flights were flown to the Philippines. In May the squadron moved to Tarakan, once that island had been secured by Allied forces. However, it took some time to set up the new camp, limiting operations to training flights and fighter patrols at dusk. Searches for missing aircraft and for Japanese suicide craft were also carried out. During June the squadron returned to Morotai and then back to Tarakan to support the Balikpapan landings.

By this time, operations were winding down. Travel flights between Morotai, Tarakan and Labuan were undertaken along with the dropping of leaflets. With the cessation of hostilities, reconnaissance flights over POW camps were initiated. During 1945 a further seven Beaufighters were lost, all but two due to enemy action.

During October 1945 No. 30 Squadron was advised it would be returning to Australia and aircraft started to be flown to Wagga Wagga. In November the squadron began moving to Deniliquin, NSW, where it disbanded in August 1946.

No. 30 Squadron used all the different marks of Beaufighter in use in Australia. The squadron's initial aircraft were in green and brown disruptive camouflage schemes, some in RAF shades of Dark Green and Dark Earth and some in RAAF shades of Foliage Green and Earth Brown. However, most probably had Sky Blue undersurfaces. Further aircraft are thought to have been received in the RAF Coastal Command camouflage scheme in which they were delivered from the UK.

Profile 43 Beaufighter Mk Ic A19-27, H, *Time Gentlemen Please / Pandemonium*

A19-27 served with No. 30 Squadron from April 1943 for just two months before it crashed on landing following the failure of the port undercarriage. The fighter was written off and converted to components.

Profile 44 Beaufighter Mk Ic A19-34, J

After receipt from the UK, this Beaufighter had a landing accident on a test flight in May 1942, necessitating repairs which took seven months. It then served with No. 30 Squadron from December 1942 until November 1943, during which time it was damaged by enemy action on three occasions.

Another of the early UK-built Beaufighters operated by No. 30 Squadron was A19-5, seen here after a landing accident at Port Moresby in November 1942 following damage from anti-aircraft fire over Lae.

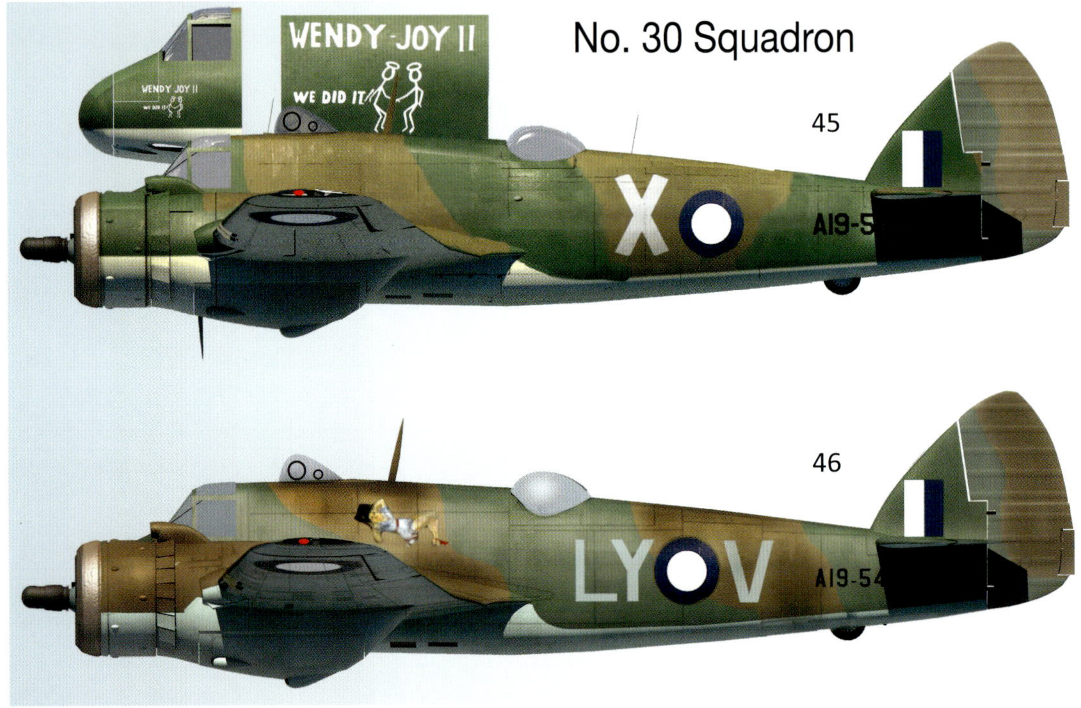

Profile 45 Beaufighter Mk Ic A19-50, X, *Wendy Joy II*

A19-50 is shown early in its service with No. 30 Squadron which it joined in August 1942. It has Dark Green and Dark Earth camouflage but appears to have had some refurbishment with RAAF colours. It participated in the Battle of the Bismarck Sea in March 1943 but the following month it was destroyed during a Japanese air raid against Port Moresby.

Profile 46 Beaufighter Mk Ic A19-54, LY-V

This Beaufighter joined No. 30 Squadron in August 1942 and enjoyed a relatively lengthy career until March 1944 when it moved to No. 5 Operational Training Unit. It appears to be in its RAF scheme of Dark Green and Dark Earth with the latter refurbished with RAAF Earth Brown.

Beaufighter A19-90 ("Z") joined No. 30 Squadron in March 1943 but had a landing accident three months later, as shown above. It was repaired and returned to service.

Profile 47 Beaufighter Mk X A19-171, LY-S

A19-171 joined No. 30 Squadron in March 1944, and it is thought to have retained its RAF colours of Dark Slate Grey and Extra Dark Sea Grey. It ditched in the sea off Noemfoor following engine failure in September that year.

Profile 48 Beaufighter Mk X A19-191, LY-C

This Beaufighter saw only brief service with No. 30 Squadron. After joining the unit in July 1944, it was damaged in a forced landing on Mapia Island the following month. It is depicted here in the RAF scheme of Dark Slate Grey and Extra Dark Sea Grey.

A19-142 ("J") served with No. 30 Squadron from August 1943 to October 1944. It is seen here early in its squadron service and appears to be in the RAF Dark Slate Grey and Extra Dark Sea Grey scheme.

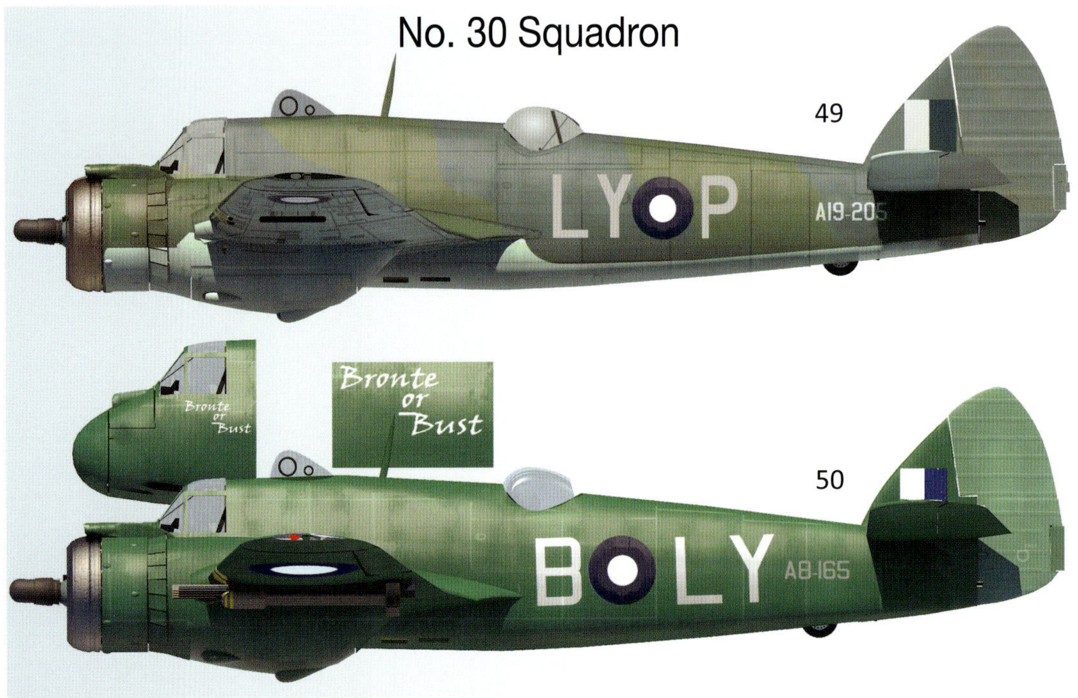

Profile 49 Beaufighter Mk X A19-205, LY-P

A19-205 served with No. 30 Squadron from July 1944 until it forced-landed the following month. It was later transferred to No. 5 Operational Training Unit at Wagga Wagga before it went into storage. It was struck off charge in August 1949.

Profile 50 Beaufighter Mk XXI A8-165, B-LY, *Bronte or Bust*

A8-165 was a Mk XXI that saw only brief service towards the end of the war. It joined No. 30 Squadron in April 1945 and six months later was flown to Wagga Wagga for storage. It was struck off charge in August 1949.

Beaufighter Mk XXI A8-154 ("Y-LY") in an overall Foliage Green scheme. It served with No. 30 Squadron from March 1945 until it was damaged in a cross wind landing at Morotai six months later.

Profile 51 Beaufighter Mk XXI A8-141, LY-K

This Beaufighter Mk XXI joined No. 30 Squadron in March 1945 and remained on strength until it was flown into storage at Wagga Wagga after the end of the war. It was struck off charge in August 1949.

Profile 52 Beaufort Mk VIII A9-488, L-LY

This Beaufort served with No. 30 Squadron from January 1945. Fitted with dual controls, it was used for training Beaufighter pilots. It had previously served with No. 100 Squadron, as shown in Profile 7 on page 39. It was flown into storage at Wagga Wagga in September 1945 and was struck off charge in August 1949.

A19-189 joined No. 30 Squadron in December 1944. In May 1945 it was transferred to No. 5 Operational Training Unit and was still coded "LY-V" when it overturned while landing on a beach north of Williamtown, as shown.

Beaufighter Mk Ic A19-84 ("EH-F"), which joined No. 31 Squadron in April 1943. Six months later it hit a tree and crashed and burned at Darwin during an interception exercise with a local Spitfire squadron. It is seen here in what is believed to be the RAF maritime camouflage scheme of Dark Slate Grey and Extra Dark Sea Grey over RAAF Sky Blue.

Seen at top is Beaufighter Mk X A19-180 ("EH-Q") that served with No. 31 Squadron from April until September 1944. The second Beaufighter is A19-155 ("EH-P") which also served with the squadron for several months in that same year.

CHAPTER 13
No. 31 Squadron

No. 31 Squadron was formed at Wagga Wagga, NSW, in August 1942. It received its Beaufighters, which were delivered from the UK, during the following two months. As the squadron was starting to work up to operational standard under the command of Wing Commander Charles Read, orders were received in October for it to move to the Northern Territory. The squadron moved to Coomalie Creek in November and commenced strafing operations over Timor within days. By the end of the month the squadron had lost two aircraft to enemy action.

Operations targeted villages and barge traffic and some missions were conducted jointly with Hudson bombers. Other operations were to protect naval vessels in the Timor Sea during which enemy aircraft were engaged. Operations continued during December with some missions being operated through Drysdale in the north of Western Australia. Attacks on Penfoei airfield in Timor were successfully carried out with a number of Japanese aircraft damaged. In December, three aircraft were lost on operations, with only one crew being rescued.

As the only Beaufighter unit in the area, No. 31 Squadron continued with a diversity of operations into 1943 including several attacks against the Japanese floatplane base at Dobo. Protective flights over convoys continued, sometimes during which enemy aircraft were driven off. Beaufighters regularly operated from forward bases at Drysdale in the west and also from Millingimbi to the east of Darwin. In the early months of the year crews experienced difficulty in some operations due to adverse wet season conditions.

The Japanese did start to fight back and during March 1943 enemy aircraft strafed Coomalie Creek and destroyed one Beaufighter. Meanwhile No. 31 Squadron strafed barges and fishing boats and attacked an enemy submarine with uncertain results. Reconnaissance flights were flown around Timor, but Beaufighter losses continued, both as a result of enemy action and accidents.

Attacks commenced on the airfield at Penfoei where four enemy aircraft were destroyed on the ground but three Beaufighters were lost. This was offset to some extent when No. 31 Squadron destroyed nine floatplanes in one attack on Taberfane. Convoy escorts continued between bombing missions as did searches for missing aircraft. New crews joined the squadron and navigation training flights became a regular occurrence.

The second half of 1943 saw attacks on the Tanimbar Islands together with regular barge sweeps. However, ten Beaufighters were destroyed or damaged due to enemy action between August and October. Although another three aircraft were lost in December, No. 31 Squadron inflicted substantial damage on the enemy with various ships, schooners and barges either sunk or damaged.

This high level of activity continued into January 1944 with cargo vessels and luggers destroyed. By this time, the Beaufighters were armed with bombs and these were being used to effect on marine targets. While these low-level attacks were producing results, aircraft continued to be lost. In addition to these operations, No. 31 Squadron was busy with various other routine tasks and on occasion the Beaufighters were also used as pathfinder aircraft for attacks by other squadrons. During July Beaufighters briefly operated from Broome. However, by this time a problem was finding meaningful targets within range of Coomalie Creek.

Armed tactical reconnaissance missions continued around Timor and adjacent islands. In August 1944, No. 31 Squadron received the necessary equipment to arm its aircraft with rocket projectiles, although this took a few weeks to eventuate. During September, three further aircraft were lost on operations, but the squadron started to receive Australian made Beaufighters as replacements. October saw a drop in activity due to a projected move to Noemfoor Island but this was deferred and limited operations against Timor recommenced and continued into November. At this time rocket attacks were undertaken, which were the first by the RAAF in the South West Pacific.

At the end of November, No. 31 Squadron moved to Noemfoor Island and commenced limited operations against targets of opportunity utilising bombs, rockets and strafing. In January 1945, the squadron began using napalm bombs and depth charges in attacks on enemy positions. Immediate use showed the effectiveness of both of these weapons. Napalm was used to destroy a powerhouse and a village as they could be dropped accurately from low level. Depth charges, dropped from a greater height, were used against lightly constructed buildings and positions where the blast effect proved most successful. Operations were conducted mainly in the Celebes area, over both land and sea, and included leaflet dropping.

During the first three months of 1945 ten Beaufighters were lost or badly damaged due to enemy action or accidents. Poor weather was a contributing factor, and all aircraft were briefly grounded pending technical investigations. Armed reconnaissance flights and sweeps then resumed around Ambon and Ceram where enemy aircraft were destroyed on the ground. During May the squadron moved from Morotai to Tarakan in Borneo, where operations began to support Allied landings in Brunei. A detachment operated from Labuan, from where No. 31 Squadron flew its last mission against installations at Kuching.

Following the cessation of hostilities, No. 31 Squadron dropped leaflets and undertook reconnaissance flights until it commenced its move back to Australia in December 1945. It was based first at Deniliquin before moving to Williamtown where the squadron disbanded in July 1946.

No. 31 Squadron operated all the different marks of Beaufighter used by the RAAF. Because its aircraft flew many of their operations over the sea, the squadron pilots preferred to retain the RAF Coastal Command camouflage scheme of the UK-built aircraft. Later, Australian-built Beaufighters operated by the squadron remained in Foliage Green.

Profile 53 Beaufighter Mk Ic A19-18, EH-T

A19-18 was an early Mk Ic delivered to Australia in June 1942. After service with No. 5 Operational Training Unit, it joined No. 31 Squadron in June 1943. In August it force landed near Millingimbi following engine failure and two months later it belly landed near RAAF Darwin due to fuel exhaustion. It was then converted to components.

Profile 54 Beaufighter Mk VIc A19-88, EH-B

This Beaufighter joined No. 31 Squadron in October 1943, nine months after its arrival in Australia and following an accident at Wagga Wagga. In February 1944 it belly landed in a sand pit near Drysdale, Western Australia, due to fuel shortage. It was subsequently converted to components.

Profile 55 Beaufighter Mk VIc A19-119, EH-W

A19-119 joined No. 31 Squadron in May 1943. The following August it did not return from a strike on the Japanese seaplane base at Taberfane in the Aru Islands. It was thought to have been shot down by two Rufe floatplanes.

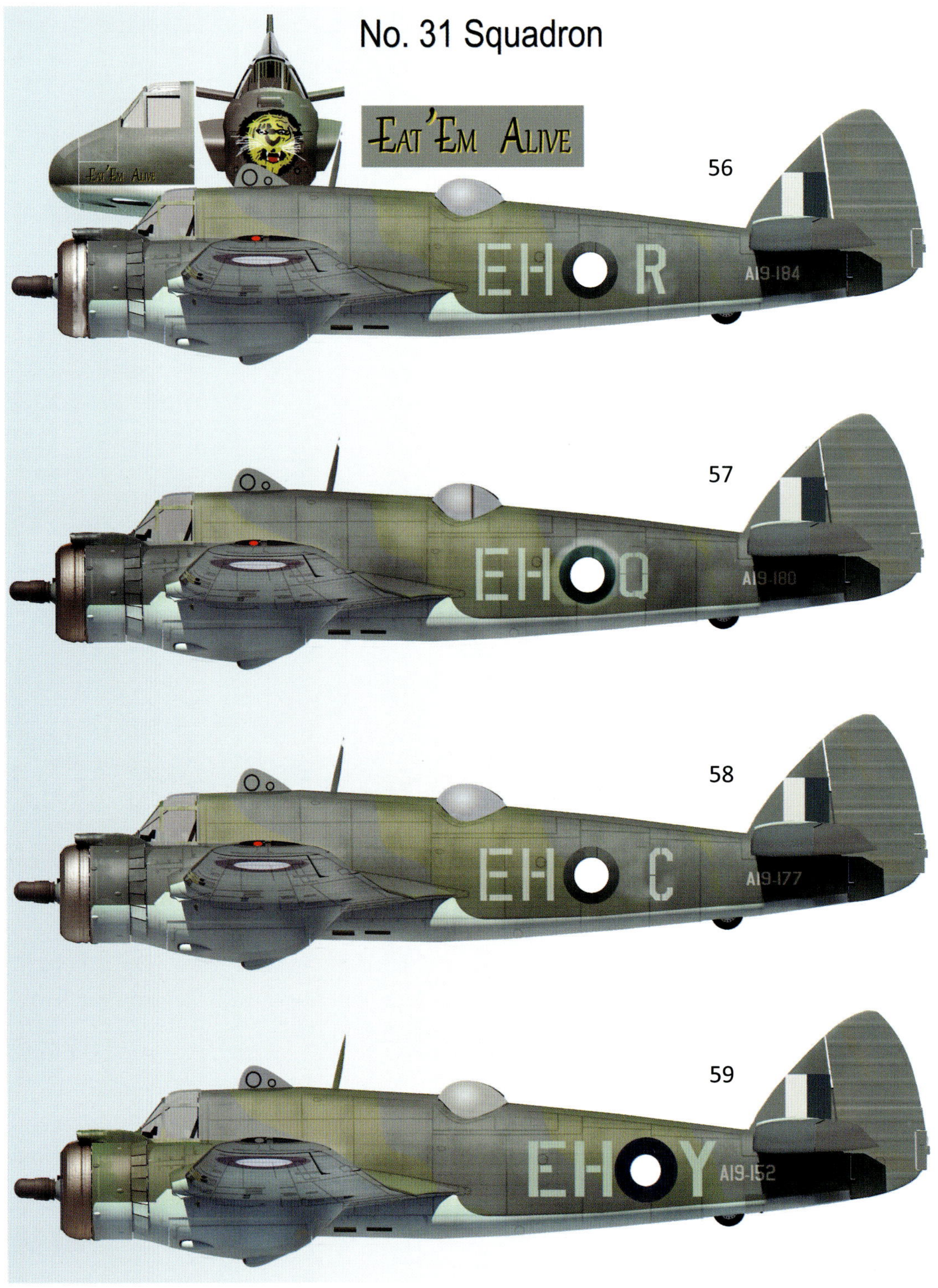
No. 31 Squadron
Eat 'Em Alive
Eat 'Em Alive
56
EH R
A19-184
57
EH Q
A19-180
58
EH C
A19-177
59
EH Y
A19-152

Profile 56 Beaufighter Mk X A19-184, EH-R, *Eat 'Em Alive*

This Beaufighter served with No. 31 Squadron from May until October 1944. After a major service, it joined No. 5 Operational Training Unit until it was stored at Wagga Wagga in November 1945. It was eventually struck off charge in 1949.

Profile 57 Beaufighter Mk X A19-180, EH-Q

A19-180 saw six months of service with No. 31 Squadron from April until September 1944. It was later used by No. 5 Operational Training Unit until it entered storage at Wagga Wagga in November 1945. It was eventually struck off charge in 1949.

Profile 58 Beaufighter Mk X A19-177, EH-C

This Mk X joined No. 31 Squadron in March 1944 but the following month it crash landed at Coomalie Creek after hitting an eagle and landing in a strong cross wind. It was written off and converted to components.

Profile 59 Beaufighter Mk XIc A19-152, EH-Y

This aircraft served with No. 31 Squadron from December 1943 to June 1944. It was subsequently operated by No. 5 Operational Training Unit until it was stored at Wagga Wagga in November 1945. It was eventually disposed of in 1949.

The two servicemen in this photo give an indication as to the large size of the Beaufighter. The aircraft is Mk XXI A8-109 ("EH-I") that served with No. 31 Squadron for just a short time from May 1945 before it was stored at Wagga Wagga later that year. It is shown here in overall Foliage Green.

Profile 60 Beaufighter Mk XXI A8-68, EH-Q, *NARROMINE III*

This Mk XXI joined No. 31 Squadron in January 1945 but was damaged in a taxiing accident at Morotai in July. Two months later it crash landed at the same location and was converted to components.

Profile 61 Beaufighter Mk XXI, A8-15, EH-P, *HOLBROOK*

This Beaufighter served with No. 31 Squadron from October 1944 through to the cessation of hostilities. In October 1945 it was placed in storage at Wagga Wagga where it remained until it was disposed of in 1949. The squadron's tiger head motif on the nose is prominent. The placement of the entire three letter squadron code to the rear of the roundel partially conceals the serial number.

This Mk XXI, A8-196 ("EH-D"), is another Beaufighter that was operated by No. 31 Squadron for just a few months from June 1945 before it went into storage in Australia. It is unusual in that it has been stripped back to natural metal finish.

A19-140 ("EH-W") joined No. 31 Squadron in September 1943. In November of that year Squadron Leader Butch Gordon was flying this aircraft when he shot down a Rufe floatplane over the Aru Islands.

Seen at Townsville is Beaufighter Mk XXI A8-43, probably during its delivery flight north to join No. 22 Squadron in December 1944. Two months later it ditched into the sea south of Morotai following an engine failure.

Beaufighter Mk XXI A8-49 that was received by No. 22 Squadron in January 1945. It seen here a short time afterwards following an accident when the starboard tyre burst on take-off at Noemfoor. The aircraft was damaged beyond repair.

CHAPTER 14

No. 22 Squadron

No. 22 Squadron was formed at Richmond, NSW, in 1936. Following the outbreak of war in September 1939, the squadron received Wirraway and Anson aircraft with which it conducted patrols along the NSW coast. In April 1942 it started to re-equip with Bostons and by October it was operating from Port Moresby and conducting regular low-level strikes against Japanese troop positions and airfields. These operations continued from bases at Goodenough Island, Kiriwina and Noemfoor Island until the squadron moved to Morotai in November 1944.

No. 22 Squadron was not long based on Morotai when a Japanese attack destroyed four Bostons and damaged another eight. As a consequence, the unit would re-equip with Beaufighters, and these began arriving in December. Training with these new aircraft was immediately instigated under the command of Wing Commander Colin Woodman and operations commenced in February 1945 with the strafing and bombing of targets in Ceram. During these operations, aircraft were damaged in low-level attacks and aircrew were injured. Sweeps and armed reconnaissance flights were conducted with attacks on targets of opportunity, including buildings and bridges. Leaflet dropping was also undertaken.

In April 1945 the squadron had its first Beaufighter fatality following a crash. On 23 April an accidental bomb explosion from a Kittyhawk injured three squadron maintenance personnel, one of whom later died. Anti-shipping strikes were carried out together with searches for missing aircraft. In May 1945 the squadron moved to Tarakan, but operations didn't recommence until June as there were delays in establishing a camp and related facilities. A detachment was based on Tawi Tawi Island in the southern Philippines at this time. The renewed operational flights included sweeps around Brunei and Labuan and throughout north Borneo.

In the middle of June, a Beaufighter crashed on take-off with the pilot being killed. Bombing and strafing flights continued but operations gradually reduced from July into August. Further aircraft were damaged in take-off accidents but only with minor aircrew injuries during August. Following the cessation of hostilities, reconnaissance flights and aircraft escort duties were undertaken through September, including throughout the Celebes, Minado and Ambon areas. During September, the squadron was mainly engaged in the preparation for returning its Beaufighters to Australia.

In September No. 22 Squadron personnel were proceeding to Morotai when their transport aircraft ditched into the sea some 60 miles east of Tarakan. Fortunately, all ten personnel were rescued without serious injury. It appears the squadron's aircraft returned to Australia for storage in October 1945, while the rest of the squadron didn't arrive at Deniliquin, NSW, until two months later. No. 22 Squadron disbanded there in August 1946.

All Beaufighters operated by No. 22 Squadron were Mk XXI aircraft painted in overall Foliage Green.

No. 22 Squadron
62
DU- A
A8-27
63
DU-P
A8-52
64
DU-Q
A8-67
65
DU-K
A8-55

Profile 62 Beaufighter Mk XXI A8-27, DU-A, *Rockabye Baby*

This aircraft joined No. 22 Squadron in December 1944 and served until the cessation of hostilities. It was flown to Wagga Wagga in October 1945 where it was put into storage until it was disposed of in August 1949.

Profile 63 Beaufighter Mk XXI A8-52, DU-P, *Pu*

This Mk XXI was operated by No. 22 Squadron from February 1945 until the cessation of hostilities. It was then stored at Wagga Wagga until it was struck off charge in 1949.

Profile 64 Beaufighter Mk XXI A8-67, DU-Q

This Beaufighter joined No. 22 Squadron in January 1945. Then during April, it was damaged twice, firstly by bomb fragments after a bomb was accidently dropped by a Kittyhawk on take-off and then the mainplane and wing tip were extensively damaged by a taxiing C-47. After several months of repairs A8-67 was allotted to No. 93 Squadron. In November 1945 it went into storage at Wagga Wagga until it was struck off charge in August 1949.

Profile 65 Beaufighter Mk XXI A8-55, DU-K

A8-55 served with No. 22 Squadron for six months from January until June 1945 when an engine failed on take-off from Sanga Sanga airstrip, Tawi Tawi. This caused it to swing to the side of the runway where it struck three parked Kittyhawks and was destroyed by fire.

The "R-DU" fuselage code on this overall Foliage Green No. 22 Squadron Beaufighter obscures the serial number, which is either A8-69 or A8-99. Both aircraft joined the squadron in January 1945.

Beaufighter Mk XXI A8-173 ("Y-SK") joined No. 93 Squadron in March 1945 but crashed on take-off from Labuan that November.

This view of Beaufighter Mk XXI A8-182 ("U-SK") shows the under-wing rockets. Named Betty, this fighter did not join No. 93 Squadron until August 1945 and so barely saw war service. By the end of that year, it was in storage at Wagga Wagga.

CHAPTER 15
The Late Beaufighter Squadrons

This chapter summarises two operational units, Nos. 92 and 93 Squadrons, that operated Beaufighters during 1945-46.

No. 93 Squadron was established at Kingaroy, Queensland, in January 1945 under the command of Squadron Leader Don Gulliver. Personnel were rapidly posted into the squadron and five Beaufighters had been received by the end of that month. Further aircraft were received during February so that by the middle of that month formation flying exercises had commenced and rocket and gunnery exercises were carried out.

During March 1945 six Beaufighters participated in army cooperation exercises with general training activities continuing. The squadron had an operational task when three aircraft were assigned to escort fifteen No. 79 Squadron Spitfires from Oakey, Queensland, to Morotai. Some accidents occurred and during April one Beaufighter was written off in a crash landing. At end of that month ground crews began moving to Morotai with the main party departing during May. One Beaufort was allotted to the squadron for pilot training duties.

In June 1945 No. 93 Squadron left Morotai and arrived at Labuan in Borneo. However, it took four weeks before the squadron's first operational mission was carried out on 26 July by a pair of Beaufighters strafing targets in cooperation with a No. 1 Squadron Mosquito. Most of the remaining squadron aircraft arrived at Labuan at this time, although one was written off in a landing accident at Townsville. Further armed reconnaissances and strafing attacks commenced.

In the two weeks of August leading up to the Japanese surrender, rocket projectile attacks were undertaken on vessels and landing grounds. This was followed by numerous leaflet dropping flights, searches for missing aircraft and photographic flights. Other missions included message dropping and aircraft escort duties. In September and October, various mundane tasks continued including leaflet dropping and mail delivery. More interesting activities included full squadron flights as a show of strength throughout the area and the sinking of surplus US barges by rockets and gunfire.

In October aircraft from the squadron started to return to Australia and in doing so escorted No. 457 Squadron Spitfires. For those aircraft remaining, weather reconnaissance flights and searches for missing aircraft continued with the squadron now down to twelve Beaufighters and half of its normal personnel. Tragedy struck in December when one aircraft was destroyed in an accident and six men were killed. The same month, the unit moved to Narromine, NSW, but it remained active in the immediate postwar period. From February to April 1946 Beaufighters escorted Nos. 76, 77 and 82 Squadron Mustangs to Japan and remained there undertaking various duties. However, by May aircraft were allotted to storage and No. 93 Squadron disbanded in August 1946.

No. 92 Squadron was formed at Kingaroy in May 1945 under the command of Squadron Leader Henry Foot, but its existence was short-lived. Personnel began to be posted in during June with Beaufighters starting to arrive on 8 July and a Beaufort a week later. Training commenced but was then cut short with the cessation of hostilities. Numerous flights continued and the most significant incident was when a Beaufighter on a joy flight crashed with the loss of all seven aboard. The squadron began winding down and by November was down to a third of its strength. It disbanded in January 1946.

No. 93 Squadron decorated both sides of the rudder of several of its Beaufighters with characterisations from the period. Many aircraft also had a name, written in script form, under the cockpit on both sides of the aircraft. Some representations are included in Profiles 66 to 70. There are no known photos of No. 92 Squadron aircraft and therefore it is not possible to accurately profile one. All Beaufighters operated by these two squadrons were Mk XXI aircraft painted in overall Foliage Green.

Profile 66 Beaufighter Mk XXI A8-159, E-SK, *Nancy*

This aircraft joined No. 93 Squadron in March 1945. It returned to Australia at the end of that year and was stored at Laverton in January 1946. It was struck off charge in August 1949. The rudder motif is the Australian cartoon character Ginger Meggs on a rocket.

Profile 67 Beaufighter Mk XXI A8-116, N-SK, *Babs*

This Mk XXI was received by No. 93 Squadron in February 1945 with the "pistol pakin' gremlin" artwork on the rudder. It served the squadron throughout that year and was placed in storage at Laverton in February 1946. It remained there until it was disposed of in 1949.

Profile 68 Beaufighter Mk XXI A8-120, P-SK, *Pattie*

A8-120 joined No. 93 Squadron in February 1945 with artwork of a bird on a rocket looking through binoculars. Underneath the artwork is the inscription *Barren Joey*. The aircraft returned to Australia at the end of 1945 for storage at Laverton. Like so many others it was struck off charge in August 1949.

Profile 69 Beaufighter Mk XXI A8-184, O-SK

This Mk XXI served with No. 93 Squadron from April 1945. In December of that year it swung on take-off from Labuan and crashed into parked Mustangs. The aircraft exploded killing the crew and four passengers. The rudder characterisation is a bee on a rocket with the words "Ye old Bumble Bee".

No. 93 Squadron
66
Nancy
EOSK
A8-159
67
Babs
NOSK
A8-116
68
BARREN-JOEY
Pattie
POSK
A8-120
69
"YE OLD BUMBLE BEE"
OOSK
A8-184

Profile 70 Beaufighter Mk XXI A8-124, T-SK, *Marge*

A8-124 joined No. 93 Squadron in February 1945. After several months of service in late August it crashed on landing at Labuan and was converted to components. The rudder characterisation is Donald Duck on a rocket.

Profile 71 Beaufighter Mk XXI A8-123, S-SK, *Paddy*

This Mk XXI was one of the first Beaufighters received by No. 93 Squadron in January 1945 and participated in the unit's first strike mission. It remained with the squadron until stored at Wagga Wagga in November 1945. It was struck off charge in August 1949.

This Beaufighter Mk XXI, A8-195 ("SK-J"), was a very late addition to No. 93 Squadron and only joined it in October 1945. It is seen after it overran the strip at Tarakan the following month where it was written off.

After service with No. 22 Squadron and repairs following damage from bomb fragments, A8-149 ("V-SK") was allotted to No. 93 Squadron in September 1945. However, its service was short lived and it was flown to Laverton for storage in January 1946.

This Beaufort originally carried RAF serial T9552 while serving with the RAF training flight based at Richmond in early 1942. It then became A9-13 and served briefly with No. 100 Squadron before joining No. 1 Operational Traning Unit in mid-1942 where it remained until January 1944. It is shown here at Bairnsdale, circa 1943, in what looks like recently refurbished Foliage Green and Earth Brown camouflage over Sky Blue.

Beaufighter Mk Ic A19-10 is shown here serving with No. 5 Operational Training Unit at Williamtown in overall natural metal finish apart from the engine nacelles, which appear to be Foliage Green. It joined this unit in January 1944 having previously served with No. 30 Squadron with which it participated in the Battle of the Bismarck Sea.

CHAPTER 16
Training Units

This chapter details three training units that operated Beauforts and Beaufighters: Nos. 1, 5 and 6 Operational Training Units, including some predecessor organisations in respect to No. 6 OTU.

No. 1 Operational Training Unit was formed in December 1941 at Nhill, Victoria, to provide advanced operational flying and instruction for general reconnaissance squadrons. It was to be equipped with several different aircraft types. Beauforts started to be received the following month together with personnel from No. 100 Squadron, RAF, arriving for training. However, they soon took their Beauforts to Point Cook and Beauforts were not on strength again until April, by which time the No. 1 OTU had moved to West Sale, Victoria.

The first courses conducted were Beaufort Conversion Courses, which commenced in May 1942. These were followed soon after by Beaufighter Courses, not with Beaufighter aircraft but instead training crews on the Beaufort to enable conversion to the single pilot seat Beaufighter. During this early period of the war No. 1 OTU crews also assisted with reconnaissance patrols to seaward.

In June 1942 No. 1 OTU moved from West Sale to nearby Bairnsdale and by this time 32 Beauforts were on strength and the unit was commanded by Wing Commander Charles Candy. The first Beaufort General Reconnaissance training course commenced in July 1942 with each course lasting between two and three months with about twenty crews in each course. Both the conversion and general reconnaissance course types were conducted concurrently. By the end of 1942 the number of Beauforts on strength had risen to 50.

In February 1943, an enemy submarine scare resulted in a detached flight of No. 1 OTU Hudson and Beaufort aircraft being flown to Laverton to participate in convoy escort duty over several days. This was to become a regular occurrence over the next few years. In April 1943, the unit assisted in searches for survivors of a torpedoed vessel and for missing aircraft. In that same month, the unit moved again, this time to East Sale. At both Bairnsdale and East Sale, personnel from the unit developed a strong bond with the local communities. This included responding to community emergencies, such as derailments and fighting bushfires, and participating in local sporting activities.

During 1944 No. 1 OTU continued to be called upon for convoy escort duty, particularly where troopships were involved. By the end of that year, the changing nature of Beaufort operations resulted in an additional course commencing, that of Beaufort Transport, to cater for the increasing use of the Beaufort in air supply. The unit was now a significant training organisation with more than 2,500 personnel. In December 1944, the appearance of a German submarine off the southern coast meant No. 1 OTU aircraft were tasked with conducting emergency anti-submarine patrols.

In both 1943 and 1944, instructional staff from the unit flew to northern operational areas to gather information for possible changes to the training syllabus. During the first half of 1945, a more significant tour of northern operational areas was undertaken. This resulted in a proposal for significant change. Operational Beaufort squadrons in New Guinea were now involved predominately in army support and no practical general reconnaissance operations had been conducted since the end of 1944. As such, the three-month general reconnaissance course being taught by the unit was becoming irrelevant. A revised syllabus was being prepared but the cessation of hostilities came before it could be implemented. During August 1945 No. 45 Beaufort General Reconnaissance, No. 30 Beaufort Transport and No. 30 Twin Engined Attack Courses had commenced but by the end of the month all training had ceased. Since formation No. 1 OTU had trained over 3,150 pupils.

A significant problem during the life of No. 1 OTU was the number of aircraft accidents, particularly involving Beauforts. These accidents were of enough concern to warrant an investigation but no one single factor could be identified. During the No. 1 OTU Beaufort training program, there were 26 fatal accidents resulting in the death of 90 men. In addition, there were 68 non-fatal accidents. In 1944 a concerted effort was made to improve Beaufort maintenance, and this had some positive effect but by the end of August 1945 two thirds of all aircraft types at the unit were unserviceable.

All No. 1 OTU Beauforts were flown to Wagga Wagga for storage starting in October 1945 with the unit disbanding around the end of that year.

No. 5 Operational Training Unit was formed at Wagga Wagga in October 1942 to train complete replacement crews for Beaufighter squadrons. The unit was initially allotted five Beaufighters and two Beauforts, the latter for pilot training being an aircraft similar to the single pilot Beaufighter. Courses commenced immediately. To assist in developing a relevant syllabus, a tour of northern areas where the two Beaufighter squadrons were operational was undertaken over the Christmas period. Squadron Leader Bruce Rose was initially in command but was soon replaced by Wing Commander David Colquhoun.

Initially one course per month was conducted and training included low-level flying with a number of accidents occurring. In the first six months of 1943, there were four accidents but fortunately only one fatality. Then during July a Beaufighter and a Beaufort were lost. By this time, the unit was operating eight Beaufighters and nine Beauforts. In August two Beauforts and two Beaufighters were lost but fortunately with no fatalities.

During October 1943 No. 5 OTU moved to Tocumwal, NSW, and training was little interrupted. However, during the Christmas period two Beauforts and one Beaufighter were lost with one fatality. By this time, there were twelve Beaufighters and twelve Beauforts available with other support aircraft also allotted to the unit, including Mosquito aircraft which had just been introduced into RAAF service. By March 1944, additional Beaufighters and Beauforts had been received although further loses in both aircraft and personnel occurred.

In July 1944 No. 5 OTU moved from Tocumwal to Williamtown, but significant losses continued, with a further four Beaufighters lost before the end of the year with seven fatalities. The new year of 1945 saw the number of Beaufighters on strength increased to 42 which enabled an increase in night flying exercises. From April to July bad weather significantly interfered with training. With the formation of new Beaufighter squadrons and the intent to convert other squadrons from Beauforts to Beaufighters, training continued right up to the cessation of hostilities in August. However, training then ceased abruptly with 29 Beaufighter courses having been conducted by No. 5 OTU. During 1945 losses amounted to ten Beaufighters with eleven fatalities.

In September, aircraft began to be sent to Tocumwal for storage and No. 5 OTU was disbanded in January 1946.

On 1 August 1942 a Base Torpedo Unit was formed at Nowra to consolidate training facilities for torpedo operations. The first course quickly commenced with four Beauforts and five No. 100 Squadron crews. The following month aircraft and crews from No. 7 Squadron arrived, but after some six weeks this squadron's designation was changed from torpedo to bomber and these crews left part way through their course. Further courses followed for other squadrons, with each regular course comprising twenty crews and each refresher training course about nine crews.

With a dual function to train Beaufort aircrews in torpedo dropping techniques and to be the chief RAAF torpedo technical centre, a broader organisation was required. Hence the Base Torpedo Unit was reorganised as No. 6 Operational Training Unit in June 1943 under the command of Wing Commander John Dibbs. In addition to training new crews, the unit conducted experimental work with various types of torpedoes. Camera torpedo attacks, drops with dummy torpedoes, formation flying and air-to-ground gunnery exercises were carried out. These were by both day and by night. Aircrews were encouraged to fly over the sea at very low altitude and accidents did occur. During April and May 1943 several aircraft ditched into the sea including following a well filmed mid-air collision over Jervis Bay.

A specific vessel was provided for training attack flights and use was made of large vessels passing along the coast. Over the 1943 winter, consistent bad weather severely interrupted training. Then in September, No. 6 OTU moved from Nowra to nearby Jervis Bay but there were not sufficient facilities to complete required maintenance, and some ground staff had to remain at Nowra. Courses continued to run but at a reduced pace and then during November there was no course for several weeks. By the end of that month, the unit had 21 Beauforts on strength.

In January 1944 six Beauforts were lost with a number of fatalities, and all aircraft were grounded for inspections. However, the overall need to continue with torpedo attacks by Beauforts was under review, as only one squadron was now operational in that role.

In February 1944, No. 6 OTU started to return to Nowra and the following month it was reduced in size by about half with only seven crews and twelve Beauforts making up each course. However, the decision was now made to cease Beaufort torpedo operations. Accordingly, after fourteen courses, No. 6 OTU was disbanded on 31 March 1944.

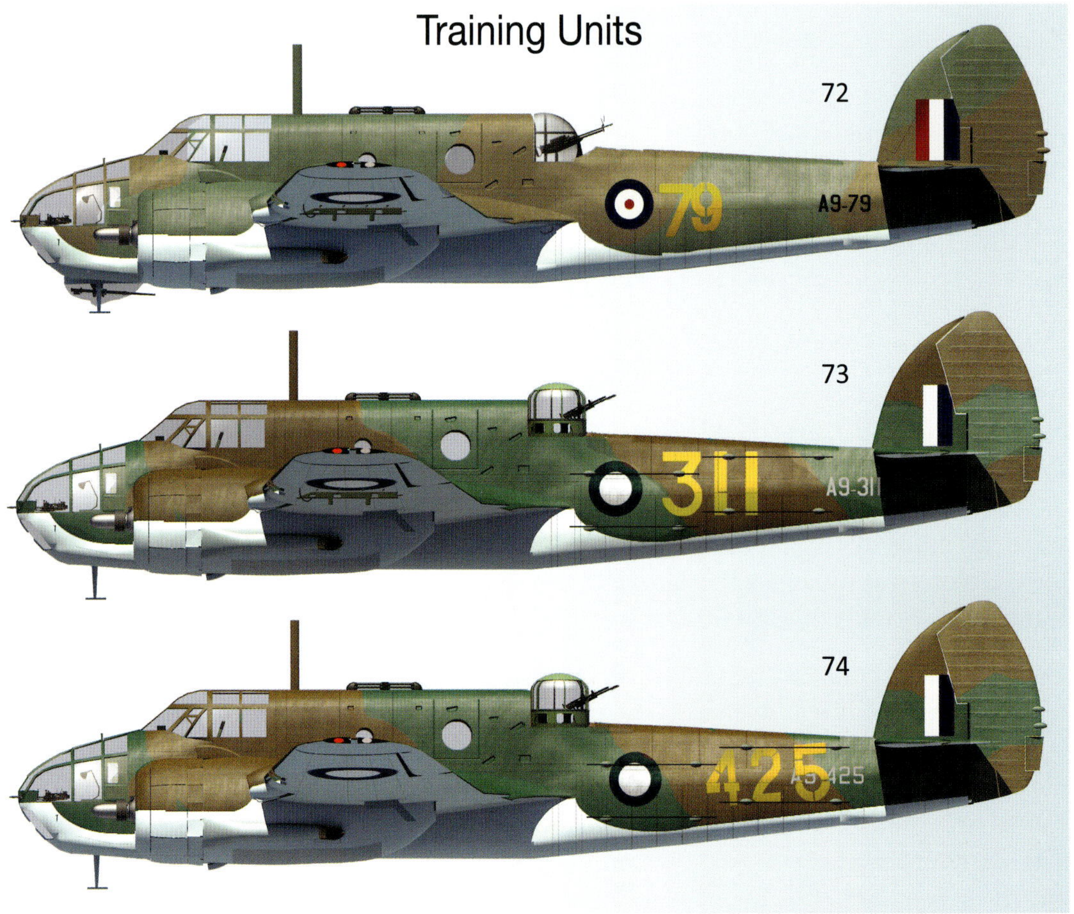

Profile 72 Beaufort Mk VI A9-79

This Beaufort temporarily served with No. 30 Squadron for the training of Beaufighter pilots before joining No. 1 Operational Training Unit in August 1942. It served with the unit as a trainer until April 1944. It was subsequently converted to a Mk IX "Beaufreighter" and renumbered A9-706. After serving with several communication units, it was converted to components in February 1946 at Morotai.

Profile 73 Beaufort Mk VIII A9-311

A9-311 joined No. 1 Operational Training Unit in June 1943. However, just three months later it crashed into hilly country north of Welshpool, Victoria, and was destroyed.

Profile 74 Beaufort Mk VIII A9-425

This Beaufort served with No. 1 Operational Training Unit from October 1943 until it was involved in a taxiing accident in April 1944 and was converted to components.

Profile 75 Beaufort Mk VIII A9-284

After sixteen months of service with No. 5 Operational Training Unit, this aircraft served with No. 1 Operational Training Unit from July 1944 until November of that year when the aircraft ground looped as it was landing at East Sale. A year later it was stored at Wagga Wagga, prior to being struck off charge in August 1949.

Profile 76 Beaufort Mk VI A9-75, G

This aircraft served with the Base Torpedo Unit and No. 6 Operational Training Unit from May 1943 until it ground looped at Nowra the following December. It was converted to components.

Profile 77 Beaufort Mk VII A9-95, J

After brief service with Nos. 100 and 7 Squadrons, this Beaufort was operated from Nowra by the Base Torpedo Unit and No. 6 Operational Training Unit for more than a year from October 1942. In January 1944 it ran off the runway into trees at Jervis Bay and was converted to components.

Training Units

78

79

80

Beaufighter A19-181 ("K") joined No. 5 Operational Training Unit in January 1945 until it crashed during a night landing that June and was converted to components. It previously served with No. 31 Squadron for seven months in 1944. It is shown here at the time of the accident in overall Foliage Green.

Profile 78 Beaufighter Mk VIc A19-126, W

A19-126 had been intended to serve with No. 30 Squadron but was found to be unsuitable due to a previous crash. In June 1944 it was allocated to No. 5 Operational Training Unit and was unusual in that its camouflage was removed back to a natural metal finish. Later that month, the aircraft force landed wheels up in a swamp near Williamtown and was converted to components.

Profile 79 Beaufighter Mk X A19-183, P

This Beaufighter served with No. 30 Squadron before it was allocated to No. 5 Operational Training Unit in April 1945. In November 1944 it had a major service and was likely repainted in the overall Foliage Green scheme, as shown here. It served with No. 5 Operational Training Unit until it was flown to Wagga Wagga for storage in late 1945. It was struck off charge in August 1949.

Profile 80 Beaufighter Mk XXI A8-80, PI

A8-80 was issued to No. 5 Operational Training Unit in March 1945 in the overall Foliage Green scheme as illustrated. The following month it swung on landing at Williamtown, damaging the port mainframe, engine and under-carriage. It was repaired and continued serving with the unit until it was stored at Wagga Wagga after the end of the war. In common with many others it was struck off charge in August 1949.

Beaufort A9-579 served with No. 1 Operational Training Unit from March 1944 until it force landed, as shown here, near East Sale just two months later. It appears to be camouflaged in Foliage Green and Earth Brown over Sky Blue.

A close-up photograph of Beaufort A9-269 that was used for torpedo performance tests at Special Duties Flight (later No. 1 Aircraft Performance Unit), Laverton, from April 1943. It was subsequently employed for various other tests including weight saving, improved elevator control, airborne photography and depth charge trials.

Beaufort A9-703 following its conversion to the Mk IX "Beaufreighter" light transport variant. It previously carried the serial A9-221, and as a Mk IX it served with No. 9 Communication Unit from February 1945 where it was fitted for spraying to control mosquitoes. It was stored at Laverton in November 1945.

CHAPTER 17
Other Units

Several Beauforts and Beaufighters served with the Central Flying School to train flying instructors in the correct manner for service flying. Beauforts with the Central Gunnery and the Air Armament and Gas Schools helped train gunnery leaders and instructors. The Department of Aircraft Production retained several Beauforts and Beaufighters for experimental work prior to passing those aircraft on for acceptance by the RAAF.

Experiments with dropping mustard gas into the jungle outside Cairns for the army were initially conducted at the beginning of 1944 using a Beaufort from No. 7 Squadron. Towards the end of 1944, an RAAF Chemical Research Unit was formed in Queensland with five Beauforts. Several research programmes were undertaken involving the dropping of bombs filled with mustard gas or the spraying of the gas.

Several of the Technical Training and Engineering Schools used grounded Beauforts and Beaufighters as instructional airframes for a number of technical functions.

No. 1 Aircraft Depot at Laverton, was a major unit undertaking a wide range of functions including modifications, major tests and trials. A number of Beauforts and Beaufighters were used for various programmes.

A major experiment that was developed at No. 1 Aircraft Performance Unit, based at No. 1 Aircraft Depot, was the fitment of DDT spraying apparatus to Beauforts for the control of malarial mosquitoes. This apparatus was used operationally throughout mosquito infested areas occupied by Allied military forces.

No. 2 Aircraft Depot at Richmond was also a major Beaufort operator, carrying out various trials, many of which were associated with radio physics.

Beauforts served with eleven separate communication units at various locations throughout Australia and the islands to the north. The Beauforts in these units undertook a variety of tasks. Certain aircraft were set aside for the use of high-ranking officers for their personal use. Many were used for courier and light freight assignments and the transport of personnel.

Three of the communication units were later renamed Local Air Supply Units and were based in Borneo, Bougainville and Tadji. As their new name implies, the Beauforts of these units were involved in supply dropping duties to troops in forward areas and to patrols of specialist forces behind enemy lines.

Beaufighters served with four separate communication units at various locations. Due to their smaller internal size, Beaufighters in these units were mainly used as fast transports for senior officers, high priority passengers and for urgently needed supplies.

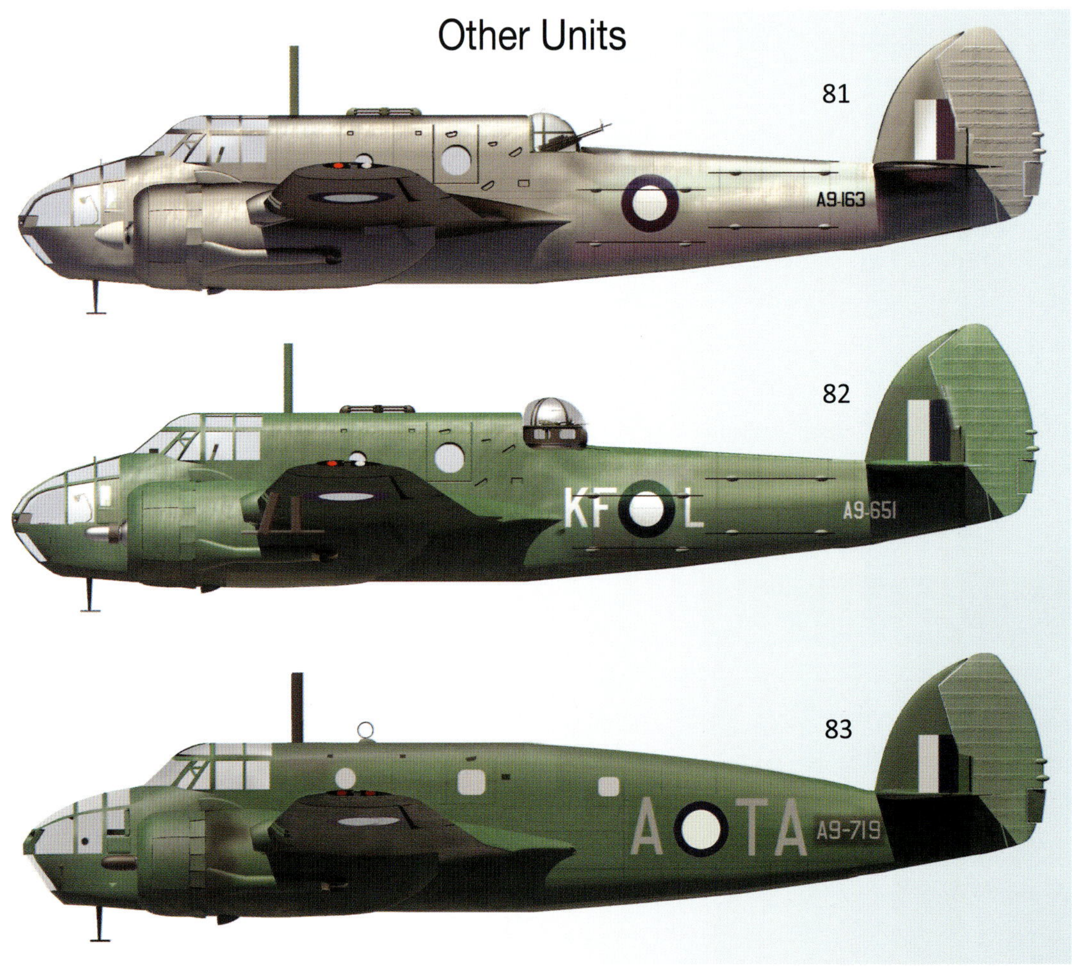

Profile 81 Beaufort Mk VA A9-163

This Beaufort served with No. 1 Aircraft Depot and No. 1 Aircraft Performance Unit for over a year from February 1943. It April 1944 it began brief service with No. 8 Communication Unit before joining No. 6 Squadron. However, in August 1944 it ran off the end of the Cape Gloucester airstrip in severe weather and was written off.

Profile 82 Beaufort Mk VIII A9-651, KF-L

A9-651 was a new aircraft when it joined No. 5 Communication Unit in August 1944. It transferred to No. 6 Communication Unit in July 1945 prior to entering storage at Wagga Wagga after the end of the war. It was struck off charge in August 1949.

Profile 83 Beaufort Mk IX A9-719, A-TA

This Mk IX "Beaufreighter" was previously A9-198. After conversion to a Mk IX, it served with No. 12 Local Air Supply Unit from June 1945 until March 1946. It was then stored at Laverton until it was struck off charge in August 1949.

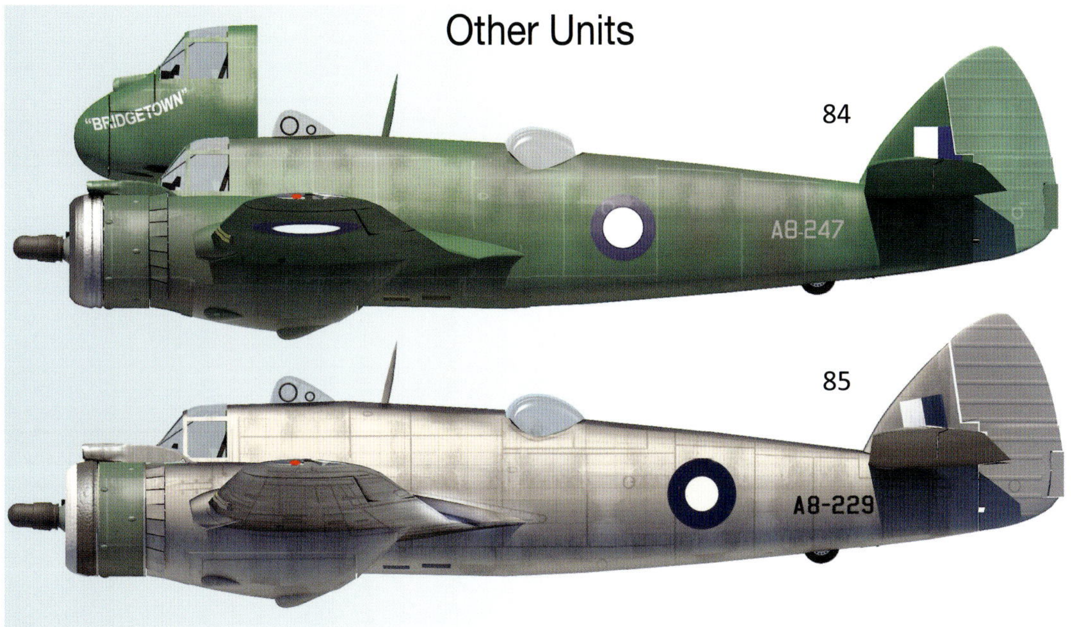

Profile 84 Beaufighter Mk XXI A8-247, *BRIDGETOWN*

A8-247, in overall Foliage Green, was not received by No. 60 Operational Base Unit on Morotai until the end of October 1945. Its service was short-lived as two weeks later it swung on take-off and crashed at Labuan and was converted to components.

Profile 85 Beaufighter Mk XXI A8-229

This Beaufighter served with No. 8 Communication Unit from June 1945 and was used for transporting senior staff officers. For this purpose, it had a natural metal finish after its camouflage was removed. From January 1946, it served with No. 111 Air Sea Rescue Flight for two months before it was placed in storage at Laverton. It was struck off charge in August 1949.

Beaufort A9-675 served with No. 1 Aircraft Performance Unit from July 1944 until March 1946. During this period, it undertook tests including on engine oil pressure fluctuations and effective range experiments with Identification – Friend or Foe (IFF) equipment.

A8-265 was used as the prototype for target tug towing in November 1945 and was used in this role from July 1946 in the Target Towing and Special Duties Flight which became No. 30 (Target Towing) Squadron two years later. It is seen at East Sale in January 1952 in overall aluminium with a yellow rear fuselage band over black and yellow striped under surfaces.

A8-363 served as a target towing aircraft from 1951, before joining the Air Trials Unit at Woomera where it is seen in 1956. It is overall aluminium with a yellow rear fuselage band over black and yellow striped under surfaces. This aircraft participated in the paravane trials mentioned in the text.

CHAPTER 18
Postwar Service

The use of the Beaufort postwar was short-lived. However, experience gained with aerial spraying became useful between 1946 and 1948 when Beauforts assisted the Department of Agriculture in spraying to protect crops from bugs and locusts.

Postwar Beaufighter service was much more significant. In August 1946 a Target Towing and Special Duties Flight was formed at Richmond equipped with five Beaufighters, one Beaufort and several other aircraft. The flight had commitments to support army and navy exercises and to assist the Council for Scientific and Industrial Research (CSIR). In 1947 one Beaufighter collected airborne dust samples for the CSIR. Also in that year, four Beaufighters were converted as target towing aircraft with a regular detachment established at Laverton, Victoria, for naval cooperation duties. Later that year a Beaufighter was used with other aircraft in a demonstration of offensive air support arranged by the School of Air Support.

Beaufighters fitted out for high altitude flying were also used over a six-month period in successful CSIR rain making experiments. Towards the end of 1947, the flight was upgraded to a Target Towing and Special Duties Squadron and then at the end of the year it was renamed the Target Towing Squadron with the special duties function now being carried out by the Aircraft Research and Development Unit.

In March 1948, the squadron was renamed No. 30 (Target Towing) Squadron and operated six Beaufighters and several other aircraft. Activities included simulated torpedo attacks and shallow dive-bombing exercises with US Navy and RAN vessels and army cooperation exercises. In 1949 the squadron moved from Richmond to nearby Schofields.

In the following years No. 30 (Target Towing) Squadron participated in rocket firing and gunnery, searchlight cooperation and cross-country exercises. Target towing exercises were also conducted for a number of other RAAF units including the Armament School, No. 78 Wing and No. 22 Squadron. Target towing was with winged targets and drogues. By 1951, servicing the Beaufighters had become a major problem due to the age of the aircraft and lack of spares.

In September 1952 the squadron moved to Canberra with the additional task of conducting radar tracking and bushfire patrols during the summer months. But this move didn't last long with the squadron returning to Richmond in April 1954. The usual activities continued until the remaining Beaufighters were flown to Tocumwal and the squadron was disbanded in March 1956. During its operations No. 30 (Target Towing) Squadron lost six Beaufighters due to accidents.

Meanwhile missile trials were being conducted at Woomera in the South Australian desert. However, the instruments contained in the missiles were being damaged when the missile returned to earth. In 1952 it was decided the missile could possibly be snatched in mid-air by being intercepted by a specially equipped aircraft. The aircraft selected was a Beaufighter, modified to tow a long cable, held out sideways from the wing by a paravane (a type of kite). Near the end of the cable was a grapnel. As a missile descended by parachute, the Beaufighter would fly across its path so that the grapnel would catch the parachute cable. Differing methods were employed to ensure a safe passage to the ground. These trials ran for four years and were commenced by the Aircraft Research and Development Unit and finalised by the Air Trials Unit.

Understandably the process took some time to perfect. After several years of trials, it was found that the Beaufighter could not carry a winch heavy enough to deal with newer missile types and it was replaced by a larger aircraft in 1956. The trials ceased shortly afterwards, with this Beaufighter the very last one in RAAF service.

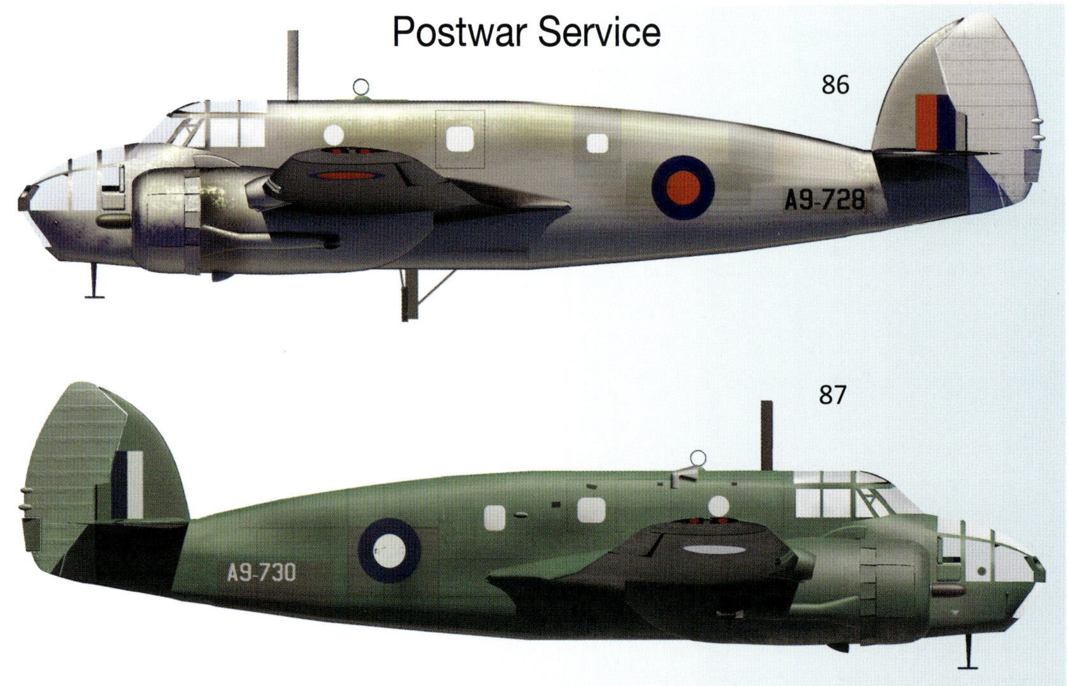

Profile 86 Beaufort Mk IX A9-728

Previously A9-213, following conversion to a Mk IX, this "Beaufreighter" joined No. 6 Communication Unit in July 1945 before it was allotted to No. 1 Aircraft Depot towards the end of that year. It was later modified for agricultural spraying with No. 1 Aircraft Performance Unit until it was stored at Point Cook in May 1948. It was struck off charge in June 1951.

Profile 87 Beaufort Mk IX A9-730

Previously A9-521, this aircraft was converted to a Mk IX in May 1945. It then served with No. 1 Aircraft Performance Unit and was modified for agricultural spraying in August 1946. In December 1946 it joined No. 1 Communication Unit and served until August 1948 when it was stored at Laverton. It was struck off charge in August 1949.

Profile 88 Beaufighter Mk XXI A8-265

A8-265 was converted to a target towing aircraft in November 1945. In 1946 it joined the Target Towing and Special Duties Flight at Richmond and continued serving with the unit when it became No. 30 (Target Towing) Squadron. In April 1954 it was transferred to the Aircraft Research and Development Unit at Laverton, but the following year it was stored at Tocumwal. It was finally struck off charge in June 1957. It is shown here in its overall yellow and black striped towing scheme.

Profile 89 Beaufighter Mk XXI A8-357

This Beaufighter was issued to the Target Towing and Special Duties Flight at Richmond and in 1947 it participated in CSIR rain making experiments. From 1951 to 1954 it served at No. 1 Aircraft Depot, Laverton. Then in June 1954 it joined the Air Trials Unit, Woomera, (later No. 1 Air Trials Unit) where it served until November 1957. It was the last operational Beaufighter in RAAF service and probably one of just two to wear the kangaroo roundel. It had a natural aluminium finish with yellow and black undersurfaces. It was struck off charge in April 1958.

Rows of stored Beauforts at Tocumwal, NSW, in 1946.

Many dozens of closely packed Beauforts and Beaufighters awaiting disposal at Wagga Wagga in the late 1940s.

CHAPTER 19
Disposals and Survivors

Following the formal surrender of Japan, Beaufort and Beaufighter squadrons based overseas began returning their aircraft to Australia. The majority went to No. 5 Aircraft Depot at Wagga Wagga and No. 7 Aircraft Depot at Tocumwal. Australian based units also began to disband and fly their aircraft to these and other bases for storage.

By January 1946, of the 700 Beauforts built, there were 376 still available for service. That month, all remaining 70 of the 180 early variants of the Beaufort were declared obsolete and these were offered for disposal. With these and other military aircraft now becoming available, the Department of Civil Aviation determined that a certificate of airworthiness would not be granted for Beaufort and Beaufighter aircraft as the aircraft were unsuitable for commercial use. However, government policy was to attempt to sell these aircraft, even if only as scrap. In April 1946, all remaining Beauforts were declared surplus to future requirements. By this time, all Beaufort units had returned their aircraft to Australia except for No. 100 Squadron which destroyed its Beauforts at Finschhafen.

The disposal of Beaufighter aircraft was similar but, unlike the Beaufort, there continued to be a reduced requirement for these aircraft in the postwar air force. By January 1946, of the 218 UK- built Beaufighters received, there were only 31 still remaining in service. That month, these 31 aircraft were declared surplus to requirements as there were sufficient Australian-built variants to satisfy all future needs. With all Australian-built Beaufighters delivered by this time there were 280 still in service. At the end of April 1946, there was an identified need for 201 Beaufighters for the post war air force. As such, 79 Beaufighters were declared surplus and were approved for disposal. The Netherlands East Indies authorities sought 50 of these Beaufighters but no sale eventuated. However, by August of that year there had been a recalculation of future Beaufighter requirements and the number was reduced from 201 to just 52. Accordingly, a further 147 Beaufighters were declared surplus and two years later another 20 airframes were added to this list.

By the end of 1948 only 10% of Beauforts and 4% of Beaufighters had been destroyed. The necessity to clear RAAF bases of all these deteriorating aircraft and free up space resulted in action. Many aircraft were sold or destroyed in subsequent months and by August 1949 only 79 Beauforts and 25 Beaufighters remained. However, disposal action was suspended because a deteriorating international situation meant it was possible the aircraft might be needed back in service. This hold notice was rescinded in July 1951 and within eighteen months most of the remaining 65 Beauforts and eleven Beaufighters at Tocumwal had been removed. Meanwhile there were only 40 Beaufighters in active service in 1953. This number gradually reduced until the last two remaining Beaufighters were sold off in 1958.

Not all of these aircraft were destroyed. Some were purchased by local farmers for parts, while complete aircraft were provided to the Department of Civil Aviation for investigations into fire prevention trials and another was set aside for the Australian War Memorial. Two Beaufighters were donated to the Melbourne Lord Mayor's Camp at Portsea, one of which was later acquired by a museum.

Several substantially complete RAAF Beauforts and Beaufighters are currently on display in museums. Beauforts are in the Australian War Memorial and the Royal Air Force Museum, while Beaufighters are at the Camden Museum of Aviation, the Australian National Aviation Museum, the Royal Air Force Museum and the National Museum of the United States Air Force.

Other examples are under active restoration. Beauforts are at the Australian National Aviation Museum, the Beaufort Restoration Group, Caboolture, and the Wings Museum, UK. Beaufighters are at The Fighter Collection, UK, and the RAAF Museum.

Spaced out rows of stored Beaufighters at Wagga Wagga, probably in 1946 or 1947.

Sources and Acknowledgements

In particular, my thanks are extended to:

Those who have contributed to and undertake the day-to-day management of the excellent resource that is the ADF Serials website.

Michael Claringbould for his patience with me in preparing the profiles and for the use of Beaufort and Beaufighter photographs from his collection.

Peter Ingman from Avonmore Books for his guidance and support in preparing this volume.

Wherever possible, the written information in this volume came from primary sources, predominately from National Archives of Australia files:

Series A705: Aircraft general instruction to C 11

Series A705: General Instruction Camouflage Schemes and Identification Markings

Series A705: Disposal of Beaufighter aircraft

Series A705: Procedure for disposal of surplus RAAF Aircraft

Series A705: Aircraft markings general technical file

Series A9186: RAAF Unit History Sheets/Operations Record Books - Forms A50 & A51

Series A9695: Early history - No 100 Squadron

Series A9845: Beaufort A9 Accidents and Beaufighter A8 & A19 Accidents

Series A10297: Aircraft status cards A9 & A19

Series A11095: [Western Area Headquarters] - Camouflage of Aircraft

Series A14487: Royal Australian Air Force Air Board Agenda

Secondary sources consulted for supporting written information include:

ADF Serials Website

Beaufort, Beaufighter and Mosquito in RAAF Service, Stewart Wilson

Beaufort Matters, Kevin Gogler

https://www.cressyaerodrome.com.au/history for Keith Parsons biographical information

Fire Across the Desert, Peter Morton

RAAF Camouflage and Markings Vols 1 and 2, Geoff Pentland

RAAF WWII in Colour series, John Bennett

Royal Australian Air Force 1939-1942, Douglas Gillson

The Beaufort File, Roger Hayward

The Identification of Various RAAF Squadron Aircraft Series, Garry Shepherson

The RAAF and the Flying Squadrons, Norman Barnes

We Never Disappoint – A History of 7 Squadron RAAF 1940-45, Kevin Gogler

Whispering Death, Neville Parnell

Photographs in this volume came from a number of sources including:

The author's Beaufort photograph collection, built over many years from a number of sources. These include the Beaufort Squadron's Association, No. 7 Squadron Association, the RAAF Museum and the South Australian Aviation Museum. Plus also photographs taken by my father.

ADF Serials Website

National Archives of Australia

State Library of South Australia

Michael Claringbould collection

No. 30 Squadron Association

No. 31 Squadron Association

Australian War Memorial

Virtual War Memorial

Destination's Journey

Pacific Wrecks

Asisbiz

Index of Names

Balmer, Wing Commander Sam 35
Campbell, Wing Commander David 67
Candy, Wing Commander Charles 101
Colquhoun, Wing Commander David 102
Dibbs, Wing Commander John 103
Foot, Squadron Leader Henry 96
Gordon, Squadron Leader Butch 89
Gulliver, Squadron Leader Don 95
Hannah, Wing Commander Colin 51
Ingram, Wing Commander Les 68
Learmonth, Wing Commander Charles 47
Newstead, Wing Commander Geoff 71
Nicholl, Wing Commander Geoff 47, 55
Parker, Wing Commander Peter 61, 62
Parsons, Air Commodore Keith 4, 41
Read, Wing Commander Charles 83
Rose, Squadron Leader Bruce 102
Walker, Wing Commander Brian 75
Woodman, Wing Commander Colin 91

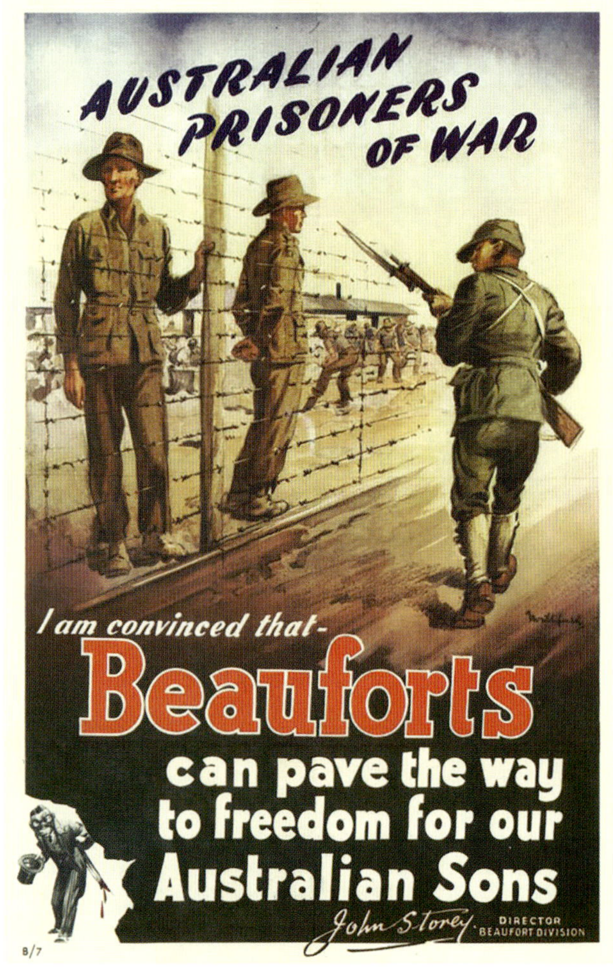

A wartime propaganda poster encouraging the Australian Beaufort production effort.